Nutrition Science

NUTRITION SCIENCE

by

Sangita Gautam

SRISHTI BOOK DISTRIBUTOR'

New Delhi 110 002

ISBN 978-81-89948-06-1

Published by :
SRISHTI BOOK DISTRIBUTORS
4378/4-B-202, 2nd floor,
J.M.D. House, Ansari Road,
Daryaganj, New Delhi-110002
E-mail : srishtibookdistributors@yahoo.co.in
Ph. 011-23284904

Edition : 2007
***Reprint* : 2012**

Rs. 900

Typeset by :
Ambey Laser Printers
Delhi.-110046

Printed at
Shivani Art Press
Delhi-110032

Preface

There are diverse concepts of nutrition. The most commonly studied aspect is the one based on laboratory research on how the diverse food components of our diet are digested, absorbed and metabolized to carry out varied activities of the body. It also assists to understand the various diseases which occur due to malnutrition and deficiency. The wealth of information encompassing these aspects of nutrition is several times more than that of food. Carbohydrate is the fuel on which the body functions are based. Energy produced through oxidation of carbohydrates forms the chief source of energy for the body. Fat is the most concentrated source of food energy, giving on a dry weight basis, more than twice the number of calories derived from protein or carbohydrate. Protein in the diet is essential to provide sources of nitrogen and amino acids to be utilized in the synthesis of body proteins and other nitrogen comprising substances. The minor components of food contribute about 2 per cent of the dry weight of our diet. Of these vitamins form the accessory food factors. Mineral salts are the second potent minor components of our diet. Some minerals in the form of their salts (sodium, calcium and potassium) are found in the body in big amounts while others like zinc, copper, cobalt, manganese and iron are present in small amounts. Some of these metals are part of several enzymes and complex organic molecules, which play a crucial metabolic role.

The process by which our food is broken down from complex insoluble substances into simple soluble ones is called digestion.

Thanks

Author

Contents

1

Diverse Aspects of Nutrition

There are diverse concepts of nutrition. The most commonly studied aspect is the one based on laboratory research on how the diverse food components of our diet are digested, absorbed and metabolized to carry out varied activities of the body. It also assists to understand the various diseases which occur due to malnutrition and deficiency. The wealth of information encompassing these aspects of nutrition is several times more than that of food.

Diet

The three basic foods which constitute our diet are carbohydrates, fats and proteins. Carbohydrates and fats are considered to be energy providing foods whereas proteins are growth-promoting substances. Mineral salts and vitamins, which are needed for several metabolic activities, form the minor components of the diet. For a diet to be adequate and balanced, these foodstuffs must be ingested in the correct proportions.

Carbohydrate is the fuel on which the body functions are based. Energy produced through oxidation of carbohydrates forms the chief source of energy for the body. Regarding quantity carbohydrates make up the bulk of the diet and consist of simple sugars such as glucose, fructose and sucrose and starch from food grains or tubers like potato. The sugar should form only a small portion of carbohydrate intake as big amounts lead to obesity and serious health problems.

Fat is the most concentrated source of food energy, giving on a dry weight basis, more than twice the number of calories derived from protein or carbohydrate. Fat is taken in lesser quantity than carbohydrates. Food fat serves as a carrier for fat-soluble vitamins A, D, E and K. Fat is stored in the body to afford energy for later utilise, its high caloric density and low solubility giving superior qualities for purposes of storage. Dietary fats may be from plant or animal sources, but, healthwise, vegetable oils containing poly-unsaturated fatty acids are preferable. The deposits of fat in adipose tissue provide both insulation and protection for the body. Barring central nervous system, virtually all tissues of the body utilize fatty acids directly as a source of energy.

Protein in the diet is essential to provide sources of nitrogen and amino acids to be utilized in the synthesis of body proteins and other nitrogen comprising substances. The proteins of the body contribute to tissue structure and are involved in a number of potent metabolic functions. Proteins in excess of these needs serve as a source of energy. The nutritional value of a protein depends upon the composition of its amino acids. Vegetable foods generally contain small quantities of protein and the amino acids available are rarely in the proportions needed by animal tissues. Therefore there is a danger of malnutrition if only one vegetable food forms the major component of the diet. However, a good vegetarian diet can be worked out which gives the complete range of protein needs by using a wide variety of protein-containing vegetable foods. These include cereals, legumes, nuts, fruits and other vegetables. Most of the animal proteins contain a high proportion of inevitable amino acids in balanced amounts and are termed biologically good proteins.

The minor components of food contribute about 2 per cent of the dry weight of our diet. Of these vitamins form the accessory food factors. These are complex organic components available in very small quantities in natural food and absorbed into the body from the small intestine. They possess no energy but are inevitable for good health and maintain the body's normal metabolic activity. If the diet is deficient in a special vitamin, metabolic activity is impaired. This produces a disorder symptomatic to that particular vitamin deficiency. This is called deficiency disease. This deficiency disease can be avoided by supplementing the diet with the missing vitamin.

Mineral salts are the second potent minor components of our diet. Some minerals in the form of their salts (sodium, calcium and potassium) are found in the body in big amounts while others-like zinc, copper, cobalt, manganese and iron are present in small amounts. Some of these metals are part of several enzymes and complex organic molecules, which play a crucial metabolic role. Our need of minerals is met by the foods we eat, epecially vegetables and leafy stuff, which constitute our diet. If need be, there could be an external supplement of minerals.

Digestion

The process by which our food is broken down from complex insoluble substances into simple soluble ones is called digestion.

Digestion starts in the mouth; when food is adequately chewed, there is thorough mixing of salivary gland secretion with food. The salivary enzyme— u-amylase—acts on the starch component of the food breaking it into shorter components - dextrins. When food enters the stomach its contents become acidic due to gastric secretion and starch digestion due to enzyme ceases. Further digestion of starch into sugar takes place in the small intestine. The mucosa of the stomach possesses several tubular glands, which are lined with diverse sorts of cells such as chief cells and parietal cells. Chief cells secrete the inactive enzyme pepsinogen and prorennin while parietal cells secrete a 0.05% solution of HC1 which makes the pH of the stomach contents 1-2.5, ideal for the optimum activity of the stomach enzymes. The acid develops the conversion of pepsinogen to pepsin which hydrolyzes protein into smaller polypeptides. Hydrochloric acid converts prorennin to rennin which in turn coagulates caseinogen, the soluble protein of milk, into the insoluble calcium salt of casein. The calcium salt is then digested by pepsin.

Comparatively little digestion of fats lakes place in the stomach since the pH of the stomach is too low for gastric lipase to be active. In the lower regions of the stomach, triacylglycerols (fats) are mixed with proteins, carbohydrates, gastric juice and other substances. Partial degradation of this mixture and churning action of the stomach led to the formation of chyme. This led to the elevation of pH, which allows some gastric lipase activity to occur. Chyme then enters the duodenum where it is blended with pancreatic juice, which comprises bile salts, pancreatic lipase and esterase. Pancreatic lipase ultimately breaks down triacylgrycerols to fatty acids and glycerol.

The pancreas is a huge gland whose cells called acini produce a number of digestive enzymes that are poured into the duodenum via the pancreatic duct. They corprise amylase to convert amylose to maltose, lipase to convert fats to fatty acids and glycerol, trypsinogen, which when changed to trypsin by enterokinase, digests proteins into smaller polypeptides, chymotrypsinogen which is converted to chymotrypsin to digest proteins to amino acids, carboxypeptidases to transform peptides to amino acids, and nucleases to convert nucleic acids to nucleotides.

Absorption

Digestion reduces food to a solution of fatty acids, glycerol, glucose, fructose, nucleotides and amino acids. If these small, soluble molecules are to be of any utilisation to the cells of the body, they must be moved out of the intestine into the blood stream. The process of moving nutrients and water from the intestine to the blood constitutes absorption.

Since the small intestine is the place where nutrient and most water absorption takes place, we should expect it to have a large exposed surface area. This is properly the case. The small intestine is very long and is packed into the abdominal cavity as a series of coils. Besides, the surface of the small intestine also contains folds and enormous numbers of fingerlike process is known as Villi. Additionally, each villus is covered with epithelial cells, and the plasma membrane of fingerlike projections is known as microvilli, it further increas as the absorptive surface area. The residues of food that cannot be digested or absorbed leave the small intestine and pass through the ileo-colonic valve into the large intestine. There, water and vitamins are absorbed, and waste materials are compacted into faces.

The absorbed nutrients do not directly go into systematic circulation. Instead, everything is channeled into the portal vein, which goes to the liver where it breaks up into tiny capillaries. The liver cells then process the absorbed food for proper utilise when they pass into the blood stream through another vein, venacava. The liver also stores absorbed and processed food for release into blood stream when needed.

The study of nutrition based on the biochemical make-up of the diet, the action of digestive juices on it, the role of the pancreas, liver, etc., has given us valuable information on diverse aspects of nutrition. The information available is voluminous. Foods, in addition to nutritionally botent components, comprise a lot of other substances in small quantities, which have significant pharmacological activity. Also cooking and food additives affect the nutritional value of food. The composition of food also ranges depending on situations of production and processing. Their effect varies from person to person who eats it. Although, the information available on nutrition is voluminous, it is responsible for some contradictions one observes in the literature.

Concern for Safeguards

Knowledge of nutrition assits man to know the types and quantities of diverse foods to be taken in our diet to maintain good health. According to ancient Indian concepts, health based on just the diet is *Arogya*—a state of no disease with which the, physician is maninly concerned. According to the World Health Organization(WHO), the term "health" means more than the presence or absence of symptoms and has a more comprehensive concept of a multidimensional, global well being.

Dimensions of Health : Growth of psychological and social sciences has shown that illness is determined by psychological, social and cultural factors. Epidemiological diseases are due to social and cultural factors, while psychosomatic diseases are because of conditions of mind. While the psychosocial concepts of health are important, they are not enough for the whole well-being of man. For, as Dr. Brisht says, "if the components of health are restricted only to physical, mental and social parameters, then a pack of wolves who are physically strong, mentally alert and socially well-knit would be ideally healthy". If this is so, a group of human beings would be no better than a pack of wolves.

Man is a multidimensional being and he has a spiritual perspective too that transcends other modalities of life. In his case the definition of health should be enlarged to cover spiritual well-being besides physical, psychological and social well-being. The Sanskrit word for health is "*Swasth*", which implies "footed in the self true health exists only when man's consciousness is fixed in spiritual self, the "*swa*".

Most persons believe that philosophy, religion, ethics and morality form spiritual health. These are only approaches to spiritual truth and not the spirit itself. When we go beyond these misconceptions and understand the spiritual reality by recognisation, we discover that there is an unity of experience that cuts across the barriers of time and space. Therefore, spiritual health conveys a special concept of a holistic and integral stand point of health. The WHO, in its programme of health for all, has recognized the spiritual dimensions of health.

Spiritual Health : Spiritual health evokes diverse images to different persons. To the materialistic Western mind it is paranormal healing, to the Eastern mind it is a boon or miracle of grace that heals our suffering and pain. To the materialist spiritual health is either a hoax or faith healing, while to —a time rationalist it is moral and ethical values in life which support man become a better race. To some others the word "spiritual" is a supreme something or someone whom none has known or can know.

According to Vedas and Upanishads (Indian scriptures), the spiritual state is a fact, which cannot be mentioned but experienced. Seekers of consciousness have come in contact with spiritual reality and their life has been moulded by its influence. They have discovered in it something rather other than our customary notions of moral, ethical and religious values. Spiritual knowledge is a state of knowledge of that, the ultimate reality; it is divulged in a few, veiled in others, and available in all.

The seers who have experienced and lived spiritual life affirm that "The spirit is other than the mind". It is unity and oneness while the mind is duality and division. It is peace and bliss, harmony and truth, while the mind dwells in pain, pleasure and indifference. The mind also fumbles through error and ignorance. The spiritual state of health cannot be defined. It is best communicated and understood in the silence and stillness of our being. It is the highest perfection man is capable of through self-evolution.

This higher and greater consciousness is not tangible to the mass of humanity as yet. The next cycle of humanity will look the evolutionary emergence of this spiritual dimension. The resurgence of interest in Yoga, the debate over the feasibility of consciousness independent of material reality, the turn towards global life, the liberalization of spiritual pursuits, the growing requirement for man for freedom, love and oneness, inclusion of the spiritual dimensions in the field of psychology, education, health and the findings in modern physics, are some areas indicative of such phenomena.

Those who have percepted spiritual heights affirm that it is feasible for man to enter and live in it and with a greater consciousness and power. If we attain that state, the disorders and conflicts that afflict humanity can be solved and it will release us from the yoke of suffering and death, and impart harmony and peace we so much lack now as a species. It is this that justifies and even makes it essential to comprise the spiritual dimension as a factor regarding to health. This factor, being a source and origin of all, will also comprise lesser terms of our existence - mental, social and even physical. The spiritual dimension is all comprehensive, the simplest and the most fundamental. It should be sought after by man who is missing peace.

Science and Spirituality : Science is related to matter. In its quest to deem matter science has acquired incredible powers and harnessed it for the objective of improving man's outer life. In the process, life has been reduced to a mechanical formula and man has been reduced to a machine among other machines.

Spirituality is the study of self and consciousness based on tapping the hidden powers of life and mind. Spirit or consciousness runs as a general thread through all phenomena. It is not an exclusive dimension that exists in isolation without any hold on the creature that inhabits it. The spiritual dimension includes the material and can intervene subtly to alter the laws and processes of the material universe and change the course of purely material and biological processes. Thus spirituality not only helps in getting spiritual dimension but also it subtly interferas to alter the laws and processes of the material universe.

Science seeks the truth of matter reality. Scientific methods and equipment can measure gross events. Scientific experiments can be seen, verified and reproduced. Scientific observations and inferences can be codified, resulting in physical laws.

Spirituality seeks after the truth of spirit, which is rather an other reality. Spiritual dimensions are subtler truth, which never be measured by conventional equipment. Subtler levels can only be experienced. One must reach a level of experience, then go to the next level and on and on till reaching an awakened state. Here, life and mind are open systems and keep growing and evolving, which upset old experiences.

Thus, science and spirituality are proceeding on their own lines of enquiry in search of a general goal. In the sincerity of their search, they will meet and when they meet they will realize that spiritual and material reality are two phases of one reality. The scientist would learn that "One Force" gives birth to the several. The spiritual scientist would discover that matter is created by a condensation of the spirit of One Reality that provides birth to various forms, phenomena and processes. This will led to in the integration of human consciousness from a fragmented one, when one experiences bliss.

Evolution of Consciousness: Man's life is riddled with disease, death and infirmity. His attempts to conquer them will remain precarious at his own level of evolution. At best,, he may arrive at a healthy equilibrium with his environment. All the emphasis and strain that besieges him is a call to evolve to a state of consciousness more perfect than his.

Man's problem of want, greed and lust and desire are due to his sense of self or individuality. His false sense 01 ego-self (I-ness) has to be shed off and replaced by true "*I*", the individual soul in him. This evolution of consciousness results in his feeling of peace, fullness and joy, with their consequences of progress, health and fitness in the physical, quietness and goodwill in the important, clear understanding and a common feeling of security and satisfaction. It will open the heart to a deep, pure and calm capacity to love without possessiveness and turbulent attachment. These will assist one to be calm and balanced, free from fear, things which assist in healthy living and even cure one of diseases.

It is feasible for consciousness to evolve further, when man will be immune from all types of diseases and even death. But, this type of consciousness is feassible only for a few. Relation with such evolved persons will help others to awaken, the sleeping soul in them, better than any other means.

2

Minerals : Its Nutrition

Calcium

The most sufficient mineral element in the body is calcium. Nearly 99 per cent of the calcium in the body is available in the skeleton. The remaining part of calcium performs a variety of functions in the body. About 850-950 gms of calcium is present in a healthy man.

Functions : The most potent function of calcium is the ossification of bones. If calcium content in the blood is below normal, bone calcium is depleted and serum level is brought to normal. Parathyroid hormone regulates calcium level in blood and its metabolism in the bone. The bones constitute a mineral reserve which may be drawn upon in requirement. Calcium is also inevitable for the formation of teeth and for the contraction of the heart and muscle. The regulation of muscle irritability is dependent upon the serum calcium level. Calcium is also needed for normal nerve impulse transmission and excitability of nerve fibres and nerve centres. Calcium forms part of acelytcholine which is essential for the transmission of nerve impulses. Calcium helps in the coagulation of blood and is required for permeability of capillary walls. Calcium exerts its influence in the absorption of vitamin B_{12} from the intestine. The mineral has been exhibited to activate certain enzymes, particularly lipase and particular proteolytic enzymes, adenosine triphosphate and rennin.

Factors Affecting Calcium Absorption : The most potent factor promoting calcium absorption is vitamin D content in the body. Utilisation of calcium is more when it is required by the body. For instance, during development the absorption of calcium is more than in adults. Calcium content in the diet enhances the absorption of calcium. Phosphates and calcium ratio in the diet also influences the absorption. If the ratio is 1:1, absorption rate is high. If phosphate ratio is high, calcium absorption is low. Phytic acid which is available in cereals combines with phosphorus and forms insoluble calcium salts and calcium phytate and inhibits calcium absorption. Abnormal intake of fats and fatty acids reduces the calcium level in blood and its metabolism in the bone. The bones constitute a mineral reserve which may be drawn upon in requirement. Minerals like iron which constitutes insoluble phosphates intervene with phosphorus absorption which indirectly affects calcium absorption. Absorption of calcium is facilitated by low intestinal pH. Normal acidity of stomach because of the presence of hydrochloric acid is inevitable to stimulate the absorption of calcium. Parathyroid hormone also aids in the absorption of calcium from the stomach. If cereals with husk are utilised in the diet, calcium absorption is reduced. Whole flours, oatmeal, and preparation with these, if not fermented, enhances the prohibition of calcium utilisation. On fermentation of the enzyme phytase splits up calcium phytate and then enhance its absorption.

Like phytic acid, oxalic acid is available abundantly in leafy vegetables and particular fruits. This fprms an insoluble compound calcium oxalate and thus diminishes the absorption of calcium. Magnesium, if taken in abundant quantities, generally ntervences with calcium absorption and results in poor calcium absorption. During sprue, steatorrhoea or in biliary fistula fat is absorbed poorly and thus collected fats help the formation of calcium soaps. Good quality protein content in the diet helps calcium absorption. During rickets citric acid increases calcium absorption.

There are several factors which hinder the absorption of calcium. Phytic acid, oxalic acid, magnesium and fat reduce the absorption of calcium. Cereals contain phytic acid which combines with calcium to form insoluble calcium phytates. Phytic acid is present in the outer parts of particular proteolytic enzymes.

Deficiency : Calcium deficiency symptoms are generally accompanied by a lack of phosphorus and vitamin D. Any of the factors which affect calcium absorption can lead to its ineffective utilisation. Vitamin D deficiency alone can produce calcium deficiency symptoms. Pure calcium deficiency is osteoperosis. In osteoperosis serum calcium level is normal but body store is reduced. In the old age the capacity of the body to absorb and uses calcium is diminished. In women hormonal imbalance during menopause or removal of ovaries at previously stages results in osteoperosis. Corticosteroid injection for any purpose may result in osteoperosis. In osteoperosis pores are seen in the bones because of decalcification. Back pain, loss of weight, fracture of bones for no reason and loss of teeth are some of the clinical symptoms. Calcium and vitamin D deficiency outcomes in rickets and osteomalacia among children and adults, respectively.

Tetany is another symptom of calcium deficiency. In tetany there is hyperexcitability of the nervous system. A series of spasmodic movements take place in the muscles. Relying upon the level of serum calcium symptoms of tetany vary. Irritation, confusion, pain, involuntary muscle spasm, bronchial spasm, spasm in the gastro-intestinal tract, pain and convulsions are again symptoms of tetany. If parathyroid gland is in hypo-activity, calcium is not properly assimilated and couses in chronic tetany. In this condition nervous changes occur and impairment of memory takes place. Depression, irritation, loss of hair and nails, roughness of the skin, pitting of teeth and epileptic fits are other symptoms of this situation. Cataract is common in such patients.

Hypercalcemia: Hyper-vitaminosis-D or Hyperpara-thyroidism produces hypercalcemia. Generally infants of 5-8 months get this problem. Loss of appetite, vomiting, constipation and flabby muscles are the symptoms.

Sources of Calcium : Milk and milk products are the best sources of calcium. Calcium is absorbed maximum from milk and milk products because of the avaitability of lysine and lactose in it. Ragi is the cheapest source of calcium; therefore it called as 'poor man's milk'. Sesame seeds and leafy vegetables are also good sources of calcium. But the availability from these sources is not insured as in sesame seeds calcium is available in its husk and on dehusking it is removed. In leafy vegetables the presence of oxalates prohibits its absorption. Molasses and vegetables such as peas, beans, potatoes, pulses, and small fish and dried fruits are rich in calcium. Unabsorbed calcium is excreted through faecal matters.

Requirements : The FAO/WHO expert group (1984) has recommended 400-500 mg of calcium per day for an adult man and woman. In the period of pregnancy and lactation the need increases to 1,000 mg/day. For infants 500-600 mg and for children of between 1 and 9 years 400 to 500 mgs and for age group 10 to 15 years 600-700 mg are recommended.

Phosphorus

No other mineral takes part in a larg number of functions as phosphorus in diverse parts of our body. An adult human body contains about 400-700 mgs of phosphorus. About eighty per cent of the body's phosphorus is deposited in the bones and teeth as calcium phosphate. The remaining portion is located as sodium phosphate and potassium phosphate in the body fluids and in cells. Phosphorus is also available in many enzymes and co-enzymes and co-carboxylase. Thus it takes part in metabolism.

Functions : Phosphorus takes an potent role in the calcification of bones. It also plays an inevitable part in carbohydrate metabolism in phosphorylation of glycogen. Phosphorus is an essential constituent of co-enzyme and co-carboxylase enzyme system in the oxidation of carbohydrate, proteins and fat metabolism. Phosphorus compounds control the storage and release of energy through ATP, ADP system. Phosphorus is present in the nucleoproteins which are an integral part of the cells. Phospholipids, which are present in diverse cells of our body, contain phosphorus. Inorganic phosphates in the body fluids act as a buffer system in the body.

About seventy per cent of ingested phosphorus is absorbed. If iron, aluminium and magnesium are present in excess amount in the diet they intervene with the absorption of phosphorus. The proportion of calcium and phosphorus is an potent factor which regulates its absorption.

Requirements **:** The daily requirement is about 1 gm. Phosphorus requirement is similar to calcium for diverse groups.

Sources **:** Animal foods like meat, fish, poultry and eggs are excellent sources of phosphorus. Milk and milk products are bigger source of phosphorus. Though cereals and legumes are good sources, phosphorus is less present from them as it is present in the form of phytic acid which is not utilised well.

Iron

Iron is a component of the compli protein haemoglobin, a haemoprotein component of the red blood cells. It is the pigment of red blood cells. Of the 3 to 5 gms of iron present in our body more than half of it is available in haemoglobin. The rest of it is accumulated in the liver, spleen, bone marrow and muscles. In the muscle it is present in the form of myoglobin. Iron is also present in some enzymes like cytochromes, catalase and peroxidase.

Functions **:** The most important function of iron is to form a constituent of haemoglobin which participate in the transportation of oxygen from the lungs to the tissues. This oxygen is utilised by the body for the processes of metabolism. In its deficiency the oxygen-carrying capacity of the blood is reduced. In red muscles myoglobin, a pigment substance, is formed with the participation of iron. This also has the same oxygen-transporting function like the haemoglobin. During strenuous activity this muscle is involved. In our body there are certain cytochrome protein pigments with iron as an inevitable component. They act as respiratory enzymes. particular flavin enzymes also contain iron in them.

Absorption of Iron **:** Dietary iron is absorbed mainly from the duodenum by the mucosa. Depending upon the requirement of iron in the body its absorption varies. If the haemoglobin concentration of the blood is satisfactory, little or no iron is absorbed and it is excreted in the faeces. Ferrous iron is absorbed more efficiently than ferric iron. In the intestine most of the iron is reduced to ferrous state. Absorption of iron is enhanced in an acid medium. Vitamin C also increases the absorption of iron, in the intestinal tract, because of its strong reducing capacity. Protein content in the diet helps the absorption of iron to some extent sulphur containing amino acids lessen iron which assists its absorption. Excess of calcium phosphate, phytates and oxalic acid interfere with iron absorption by forming insoluble iron salts or oxalates. Minute traces of copper support the formation of haemoglobin. The rate of absorption of iron is affected by the magnitude of iron store in the body and the rate of erythropoiesis. Malabsorption syndromes also reduce iron absorption.

Deficiency of Iron **:** Anaemia is a condition where the haemoglobin level is lowered in the blood. Lack of iron in the diet is the most common cause of anaemia. Iron, folic acid, proteins, B_{12}, copper, cobalt, vitamin C and pyridoxine are the nutrients required for erythropoiesis. Folic acid deficiency produces megaloblastic anaemia.

Symptoms of Anaemia **:** When the haemoglobin level falls below 12.5 gm/100 ml of blood, it results in the diminished oxygen-carrying power of the blood. Iron deficiency anaemia is otherwise called hypochromic anaemia. In this situation the number of RBC is not much reduced but the quantity of haemoglobin is less. The pallor of skin and tissues, fatigue, weakness, giddiness, blurred vision, anorexia, heartburn, palpitation, dysproea, oedema over feet, headache and a realising of tiredness are some of the symptoms. The nails show koilonychia and tongue and papillary atrophy are seen. Among women menstrual disturbances like menorrhagia or phases of amenorrhoea take place.

Nutritional megaloblastic anaemia occurs due to dietary deficiency during old age, increased need during pregnancy, lactation, prematurity, rapid growth and in endocrine disturbances. Haemolytic anaemias occur during malignancy, rheumatoid arthritis, pulmonary tuberculosis and in skin diseases. Alcoholism and side effects of drugs also result in anaemia. In diseases such as idiopathic steatorrhoea, tropical sprue and in partial gastractomy anaemia takes place. In the period of pregnancy this sort of anaemia usually occurs in the second trimester. Weakness, palpitation and fatigue are followed by fainting bouts.

Anaemia must be treated with ferrous sulphate tablets and iron-rich diet.

Hemosiderosis : It is a disorder of iron metabolism. In this situation large deposits of iron are made in the liver and in the recticulo endothelial system. Iron with carrier protein which called as transferring of circulation becomes saturated with iron. When transferrin is unable to bind all the absorbed iron this metabolic disorder of iron takes place. If iron vessels are futilised for almost all cooking; much more iron is ingested in the body, which may cause in hemosiderosis. The Bantu population of Africa is an instance of this condition.

In hemolytic anaemia a big number of red blood cells are destroyed and this can lead to hemosiderosis. Prolonged prophylaxis with big doses of iron can often produce hemosiderosis.

Sources of Iron : Liver, kidney, heart, lean meat, egg yolk and shellfish are the best sources of iron. Dried beans, dried fruits, legumes, nuts, green whole cereals, leafy vegetables, enriched grains and molasses are good sources. Milk is a poor source of iron.

Requirement of Iron: Recommended intake of iron is calculated, based on its loss and absorption. For an adult man weighing 55 kg the iron need is 24 mg/day and for a woman of 45 kg weight 32 mg is the requirement. During pregnancy 40 mg and during lactation 32 mg of iron is needed.

For infants till one year 1 mg/kg body weight and for children of 1 to 9 years of age 20 to 25 mg of iron are needed. For boys and girls between 10 and 12 years of age 30 to 25 mg and for boys 13 to 15 years of age 25 mg and for girls from 13 to 15 years 35 mg of iron are required. In the later adolescents for boys (16-18 years) 25 mg and for girls 35 mg is the recommended allowance.

Magnesium

The human body contains minute quantities of magnesium in the bones along with calcium and phosphorus and the rest in soft tissue. An adult human body has about 25 gm of magnesium.

Functions : Magnesium plays an potent role in normal calcium and phosphorus metabolism in man. It is also present in particular enzymes like co-carboxylase which decarboxylase pyruvic acid. Magnesium also takes part in the phosphorylation in metabolic pathways.

Magnesium is readily absorbed from the intestine. Excess magnesium is excreted in the faeces. Normal human serum contains about 2 to 3 mg/100 ml.

Deficiency: Prolonged starvation brings magnesium deficiency, vomiting and intestinal fistula. Diabetic acidosis is related to raised serum level of magnesium. Neuromuscular irritability, tremors, cramps, twitching and convulsions occur in its deficiency. Confusion, disorientation and visual hallucinations are some other complaints with magnesium deficiency.

Sodium

Sodium is inevitable for the normal functioning of the body. Sodium plays an potent role in the regulation of acid-base balance and water metabolism in the body. Sodium in the form of sodium chloride is ingested directly through food and several food materials contain this mineral. Almost all the body sodium is found in the extracellular fluids of our body. Through sodium the osmotic pressure is also regulated by the body. A healthy kidney regulates the excretion of sodium according to the need. In case of kidney diseases sodium is retained in the body and oedema takes place. Under normal conditions 90 per cent of ingested sodium is excreted in the urine in the form of sodium chloride and sodium phosphate. Sweat contains some amount of sodium. Sodium assist to contribute alkalibility of the gastro-intestinal tract.

Sodium plays a particular role in controlling the heartbeat by helping in its origin and maintenance. Contraction of muscles and regulating the permeability of cell membranes are the other functions of sodium.

The adult human body comprises about 100 gm of sodium. On an average diet about 3 to 5 gins of sodium is excreted in the urine. In profuse perspiration sodium chloride is depleted from the body. In sodium depletion muscular cramps, weakness, headache and vascular collapse take place. In decreased water intake the serum level of sodium is enhanced and this situation is known as hypernatraemia. Blood urea is increased and cellular water is depleted during dehydration and in hyponatraemia. In infants shrinkage of brain causes rupture of cerebral veins and further cerebral thrombosis. Retention of sodium produces oedema and ascites. In renal diseases, cardiac failure, hepatic cirrhosis with ascites, toxaemia of pregnancy, prolonged administration of corticosteroids sodium must be restricted through diet. Generally 4-5 gm sodium is injested through food and it is restricted to 250 mg 500 mg, or 750 mg depending upon the condition.

Requirements: For an adult 1-2 gms of sodium and for children 350 mg-1 gm is prescribed. For infants 150-400 mg is enough.

Sources: Animal foods like meat, meat extractives, fish, poultry, milk, milk products, nuts, bakery items, salted biscuits, egg-white, dried fruits, roots like beetroot, carrot, radish and leafy vegetables like spinach are excellent sources of sodium.

Abnormalities of Sodium Level in Blood: Hyponatraemia

and hypernatremia are two situations which manifest the sodium level in the blood.

Hyponatraemia: When serum sodium level falls below normal the condition is known as hyponatraemia; various situations lead to this symptom. Severe dehydration, decrease in blood volume, decrease in blood pressure, circulatory failure are some of them, prolonged vomiting, diarrhoea, chronic renal diseases with acidosis, Addison's disease because of adrenocestical deficiency are the common causes for hyponatraemia.

Hypematraemia : In this situation the plasma sodium iron content is higher than normal. Hyperactivity of adrenal cortex as in cushing's syndrome, prolonged treatment with cortisone, ACTH and sex hormones are some of the causes of hypematraemia. Deficient water intake because of vomiting, nausea, difficulty in swallowing with sore throat, oesophegal disease, cerebral impairment, enhanced water loss through respiration diarrhoea, excessive urination in diabetes mellitus, chronic renal diseases, enhanced sodium intake and deficient sodium excretion brings hypematraemia.

Potassium

In contrast to sodium, potassium is available in the intracellular fluids of the body. It is an important constituent of cells. In adult body 250 gm of potassium is available. Most of it, that is, 97 per cent is found in the intracellular fluids and the remaining 3 per cent in extracellular fluids. The other potent function of potassium is that it influences the cellular excitability of smooth, skeletal, cardiac and nervous tissues. If potassium content is high muscular irritability is enhanced. Diminished calcium and magnesium also result in such a condition. Involuntary muscular function is controlled by potassium content in the body. When the potassium level in the myocardium is changed it is reflected in the electrocardiogram. Potassium is also essential for the acid base balance in the body.

Potassium deficiency shows weakness in muscular functions, mental disorientation and cardiac irregularities.

Hyperpotassemia results from kidney diseases when renal excretion is less. In dehydration these symptoms occur. Deficiency of potassium is not common unless as a complication of diseases. Deficiency condition shows hypokalaemia in the period of operation when food cannot be ingested. In diabetic acidosis this condition occurs. Deficient intake because of starvation and excessive loss from gastro-intestinal tract leads to deficiency.

Hypokalaemia takes place because of deficient intake of it while a surgical patient is exposed to prolonged parent therapy, excessive loss from gastro-intestional tract through persistent vomiting, fistulae or diarrhoea before and after surgery, the patient may be on a poor diet which may increases the excretion of potassium. Surgical cases because of low intake produces hypokalaemia.

In diabetic acidosis potassium is lost through dehydration on insulin therapy, sugar is converted to glycogen and starch. This glycogen formation depletes some potassium from extracellular fluids and serum potassium level falls and it is dangerous.

While diuretics are administered considerable loss of potassium through urine leads to hypokalaemia. Nephritis and other kidney diseases, tumours of large intestine, treatment with steroids, deficiency disease like kwashiorkor, megaloblastic anaemia, heatstroke and violent exercise are some of the other reasons which produce hypokalaemia. Prolonged hypokalaemia causes injury to myocardium and kidney.

Hyperkalaemia **:** When the potassium level in the serum is high the condition is called hyperkalaemia. This occurs because of diminished urine volume, intake of potassium salts of penicillin, iodide, acidosis, endogenous tissue breakdown, Addison's disease, kidney failure due to kidney disease or chronic renal failure and rapid intravenous administration of potassium are some of the reasons for hyperkalaemia. Cardiac and the central nervous system are affected by this situation. Anorexia, weakness of the muscles, diminished tendon jerks, low heart sounds followed by peripheral vascular collapse and common apathy are some of the symptoms.

Trace Elements : Elements which occur in our body in small quantities are known as trace elements. Copper, iodine, zinc,. manganese, molybdenum, fluorine, selenium, nickel, chromium, cadmium, silicon, vanadium, strontium, cobalt and chlorine are some of the trace elements. Though these trace elements are available in traces they are very inevitable to maintain the structure of various organs in good condition. RBC formation, thyroxine production, formation of healthy bones, haemoglobin formation and proper enzymatic actions are some of the potent functions of trace elements in our body.

Iodine

The significance of iodine as an inevitable trace element lies in its role in thyroxine production. Thyroxine is the hormone produced by the thyroid gland. Iodine is the/main component of thyroxine and a normal -adult has only about 10 mg of iodine in the thyroid. The absorbed iodine combines with a thyroxine binding protein. Thyroxine plays a potent role in carbohydrate metabolism. Deficiency of thyroxine interferes with the normal physical and skeletal development of the child during the growing period. Iodine also assists the tissues in its oxygen consumption. Conversion of glycogen to glucose occurs only in its presence. Thyroxine enhances heart rate and depletes bone calcium.

Deficiency : If the iodine intake is deficient the thyroid gland enlarges and this condition is called simple or endemic goitre. The normal thyroid gland weighs about 25 gin. In goitre it enlarges from 200 to 500 gm. Iodine deficiency goitre causes in hypof unctioning of the gland. If diagnosis is done early and iodine supplements are given the thyroid may become normal. If timely treatment is not given the enlargement will become permanent. In the growing period if severe deficiency occurs it results in retarded growth and the condition is called cretinism. People living particularly in submountain zones are affected. In India residents of sub-Himalayan regions and Kashmir and parts of adjoining plains are the victims. The most affected are among girls between the ages of 12 and 18 years and boys from 9 to 13 years. The areas where goitre is common is called the goitre belt. According to the latest data available, about

40 million people are affected with goitre. In these areas air, water, soil and foods grown are poor in iodine which results in its high incidence. Goitre can be prevented by the regular use of iodized salt.

Goitrogenic Substances in Foods : Particular foods of brassica genus like cabbage, cauliflower, turnip, soyaflour and oil contain substances which react with iodine available in the food and make it unavailable. These substances are known as goitrogenic substances. Consumption of these foods in large quantities makes the iodine non-available to the body.

Requirement of Iodine: An adult needs about 0.15 to 0.2 mg of iodine and for infants and children 0.05 to 0.1 mg daily is enough.

Sources : Iodine is available in minute quantities in all foods. The iodine content of foods depends on the iodine content of the soil. In particular parts the soil contains less iodine and the foods grown in such places are iodine poor. The soil of mountainous regions is generally iodine-poor soil. The best sources of iodine are seafoods, common salt and vegetables; cereals and legumes are poor sources of iodine. Most of the iodine in the vegetables is in their leaves and flowers.

Copper

Copper is found in all tissues of our body. It is an inevitable element in the formation of haemoglobin. About 100 to 150 mg of copper is available in the human body. The brain, liver, heart and kidney contain the highest amounts. In the blood, copper is available in the plasma and RBC. Copper is absorbed rapidly into the blood.

Copper is a component to particular enzymes in our body. It is part of the enzyme cytochrome oxidase which is inevitable for glucose metabolism. It is also part of tyrosinase, an enzyme necessary for converting the amino acid tyrosine to melanin, the dark pigment of the skin. Copper stimulates the absorption of iron. Copper is a constituent of the elastic connective tissue protein elastin.

Deficiency: Copper deficiency is not common in human beings. In malnourished persons iron is not absorbed adequately and anaemia occurs. Premature babies fed on cow's milk develop anaemia due to copper and iron deficiency.

Requirement: An adult person requires about 2 mg/day. In the period of pregnancy and lactation the copper need is increased to 3 mg/day. Infants from 0 to 12 months require 0.5 mg to 1 gm and children 2 gm and adolescents 3 mg of copper.

Sources : Millets like cholam, bajra, raw rice, whole grams, cow gram, dried peas, red gram dal, vegetables like drumstick are good sources of copper. Liver is the richest source of copper. Organ meat like kidney is also a good source of copper.

Zinc

In an adult 2 to 3 gms of zinc is available. It is present in particular parts of the eye, particular glands and in their secretion. Liver, muscles, bone and hair contain some of the remaining zinc. Blood, particularly the RBC, has some zinc. In small amounts it is present in tissues and bone cells.

The potent function of zinc is its role in enzymatic action. At least in 25 enzymes of digestive and metabolic action zinc takes part. It is also present in an enzyme carbonic anhydrase which is inevitable for the transport and elimination of carbon dioxide. Co-carboxypeptidase and alkaline phosphatase contain zinc in them. Zinc plays an essential role in the formation of DNA and RNA. It is a constituent oi insulin and it is necessary for the formation of connective tissues. Zinc also aids in the healing of burns and wounds.

Deficiency: Zinc deficiency in animals causes in growth failure, loss of appetite, degeneration of the male sex glands, testes, skin lesions, alopaecia or loss of hair, changes in the epithelial and cutaneous tissues, dermatitis and scaling and cracking of the paws. Abnormal bone formation, thickening and metabolism also occur during zinc deficiency. In human beings zinc deficiency produces anaemia, development retardation and delayed genital maturation.

Requirement: Infants require 3 to 5 mg. For children 10 to 15 mg and in the periad pregnancy and lactation 20 to 25 mg of zinc are required.

Sources : Seafoods, meat poultry and eggs are good sources of zinc. Cereals, legumes, and nuts contain considerable amounts. Fruits and vegetables are poor sources. On milling eighty per cent of zinc is lost

Cobalt

In an adult body about 1.1 mg of cobalt is present. It is found in the blood, liver, kidney, spleen and pancreas. It is also available in minute quantities in all tissues.

Function : The most potent function of cobalt is that it forms an integral part of vitamin B_{12}. In animals, if the feed is deficient in cobalt severe anaemia occurs. Addition of cobalt to feed or injection of vitamin B_{12} cures this ituation. In human beings vitamin B_n synthesis occurs in the colon. At that phase absorption is not taking place. So preformed B_{12} must be ingested for the cobalt need. But deficiency of cobalt is not common among human beings. Cobalt also takes part in the enzymatic action and also in thyroid function.

Requirement: Since deficiency is not known, the available is not studied. In pernicious anaemia B_{12} deficiency is available and on administration of B_{12} it disappears.

Sources : Cereals, beans, peas and organ meats like kidney and liver are good sources.

Manganese

About 20 mg of manganese is available in an adult. Bones, liver, pancreas, kidney and pituitary gland contain manganese.

Function : The manganese functions in several enzyme systems. It activates particular enzymes which take part in the digestion and metabolism of carbohydrates, proteins and lipids. In the synthesis of ATP cholesterol, fatty acids and RNA, the manganese plays a potent role.

Deficiency : Manganese deficiency in human beings is not known. Experimental deficiency symptoms are slowing of growth, abnormalities in bone, depressed reproductive functions, neurological weakness of the limbs, ataxia and lack of balance. Among birds a situation known as perosis occurs on manganese deficiency. It is characterised by bone deformities and dislocation of bones.

Sources : Cereals, ,bran, dried beans, peas, green vegetables and nuts are good sources. Tea and coffee are high in manganese content. But animal foods are relatively poor in it.

Chromium : An adult body comprises about 6 mg of chromium. Chromium is available in skin, hair, muscles, brain, tissues, adrenal glands and in body fats.

A petent function of chromium is its role in glucose tolerance factor formation in the body. Chromium niacin complex factor assists the action of insulin on glucose. Chromium is also inevitable for the activating enzymes involved in carbohydrate, protein and fat digestion and metabolism. The serum cholesterol level was raised on experimental animals in chromium deficiency.

Requirement: The exact needs is not known. A normal balanced diet supplies enough chromium.

Sources : Organ meats like liver, meats, wholegrain cereals and brewer's yeast are good sources of chromium.

Fluorine

Fluorine is available in the body as fluoride. It is present in teeth, bones, thyroid gland, skin and liver.

Fluoride is esssential for the formation of caries-resistant teeth. Small amounts of fluorine prevent dental cariesin children. It promotes calcification.

Fluoride and Tooth Structure : For healthy tooth formation some amount of fluorine is inevitable. Though the exact mechanism by which it prevents dental caries is not fully understood, it is assumed that fluorine is incorporated into the tooth enamel. This reduces the solubility of enamel in the acids produced by bacteria. In its scarcity the enamel is soluble in the acid and tooth decays. In experimental animals fluoride seems to assist normal development during the growing period, but in human beings its role is not yet proved.

Toxicity or Fluorosis : If water contains more fluorine, dental fluorosis occurs. The enamel of the teeth loses its natural lustre and becomes rough. Opaque white Patches with yellow or brown stains are found irregularly on the teeth. Ridges or diffused spots and pitting take place over the teeth. The surface of the teeth gives a corroded appearance.

Skeletal Fluorosis : The bones of vertebral column, particularly the cervical part of the spine, Pelvis and bones of the lower extremities, are affected in this situation. Hypercalcification of the bones and the collagen take place because of excessive fluorine deposition. Stiffness and pain are the prior tymptoms. Weakness and spasticity of lower limbs, wasting of the distal muscles of hands, deep jerks on lower extremities and numbness or girdle pain are the other symptoms of skeletal fluorosis. Gradually the person becomes crippled as he cannot bend or squat for become stiff.

Requirements : Water containing 1 to 2 ppm prevents caries and so it is considered as the need.

Molybdenum

The adult human body contains only 9 mg of molybdenum. It is avaitable in the liver, kidney, adrenal glands, bones, skin and tissues. Molybdenum is essential for the normal enzymatic system and its, function. Enzymes such as xanthine oxidase and aldehyde oxidase make use of molybdenum. These enzymes are inevitable for the proper nucleic acid and iron metabolism. Molybdenum seems to aid in preventing teeth decay. The need of molybdenum is not known. Wholegrain cereals, dried beans, peas and dark green leafy vegetables are good sources. Meat is also an excellent source.

3

Vitamin D : Its Nutrition

The human requare for vitamin D except in the case of infants under one year and for expectant and nursing mothers. This uniform level is an reality a confession that very little is known about varying requirements. At one period it was expected that vitamin D was needed only during the time of active growth. This view is no longer held and it is approach that the fragile nature of the bones people of older may be concerned to their low consumption of vitamin D foods.

The Health Committee of the League of Nations made the first estimates; these were followed by American estimates. In view of the importance and difficulty of the task, it is surprising how closely the diverse estimates agree and it is correct that these recently published figures are as near the truth as we are likely to get.

The question has to be asked to what extent do individuals eat diets that satisfy these needs? How easy or problem is it to obtain a sufficient diet? There are so several ways of approaching this problem. We can see whether there is any evidence that people are better fed now than formerly or we can study foods that are normally eaten in ordinary homes and see if they supply the requirements.

Reference has been made to the effect of adding protective foods to the diet of schoolboys and to the increases in height and weight yielded as a result. Data on the height and weights of London school children during the first half of this century have recently been published and these show the same type of increase. It may not be totally due to nutrition, but it is not unreasonable to arrive to conclusion that improved feeding since of increased knowledge of nutrition has also played a role in producing these effects.

If we access the difficult from the dietary angle, we can analyze representative diets and estimate their vitamin content. This is a purposeful exercise and a comparison between the foods eaten on several separate days or by dissimilar individuals displays how wide the variation in vitamin content can be.

Historical Backdrop

Rickets in central Europe came to be known as the English disease! In England in 1900 rickets became epidemic due to industrialisation as parents got them selves engaged there leaving babies inside the door.

A combination of clinical observations of children afflicted with rickets and of experimental animals in whom rickets had been developed under controlled laboratory conditions led to the identification of vitamin D as the *antiricketic vitamin*. An early (1890) greater observation, made by palm an English physician, was that the incidence of rickets is related to the amount of sunshine available; where sunshine was abundant, rickets was rare; where the sun seldom shone, rickets abounded. After a few years the concept developed that rickets, such as scurvy and beriberi, may be a dietary deficiency disease. With this in mind, the expected next step was to produce the deficiency in experimental animals. In 1921 Mellanby in

England yielded rickets in puppies by using a mixture of natural foods; then he induced recovery by offering them cod liver oil. At that time cod liver oil was known to include only vitamin A, so the cure was attributed to it. Later it was invented that even after the vitamin A content of cod liver oil was destroyed by oxidation, the antiricketic potency still remained. It was then that it became explicit that cod liver oil contained two vitamins, vitamin A and the antiricketic factor. In the meantime, the association of rickets with calcium and phosphorus metabolism, and the favorable ascendancy of cod liver oil on the utilize of calcium were recognized.

By 1922 the mystery of the effectiveness of sunlight in preventing rickets was on the source to solution. Rats fed a rickets-producing diet failed to develop rickets if exposed to sunlight. A part from, rats fed a rickets-producing diet that had been irradiated in the laboratory, provided the same influence as exposure to the sun, also failed to develop rickets. After the discovery of the effectiveness of irradiating the diet as a whole, each item was removed, one at a time, and treated separately then incorporated into the diet in an attempt to find out which nutrient was accountable for the antiricketic effect. Fat was the only nutrient that on irradiation imparted antiricketic potency to the diet. With fats established as potential antiricketic factors, the next question to be solved was which portion of the fats, the alcohol or fatty acid fractions, became *activated* on exposure to the ultraviolet rays. It was found to be the alcohol portion. Intensive research immediately revealed that numerous alcohols, all of which belonged to the sterol group of alcohols, acquired antiricketic activity on irradiation. The sterols are much more complicated in chemical structure than the general medicinal and beverage alcohols.

By the late 1920's it had been established that rickets could be curbed and cured by exposure to direct sunlight, by irradiation with ultra-violet light, by feeding irradiated food, and by feeding cod liver oil. Then followed the recognition of the *natural* vitamin D of fish-liver oils as the same materials produced in human skin on irradiation of 7-dehydrocholesterol. The vitamin D formed from 7-dehyrocholesterol is designated as vitamin D_3. The irradiation of ergosterol, a provitamin D in plant foods, produces a dissimilar substance, designated as vitamin D_2.

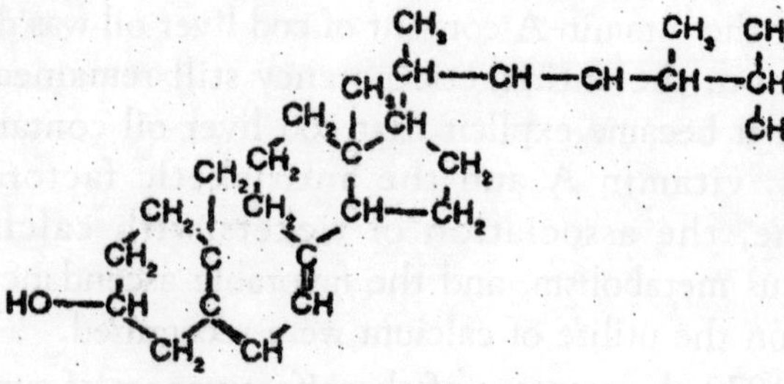

Vitamin D_2

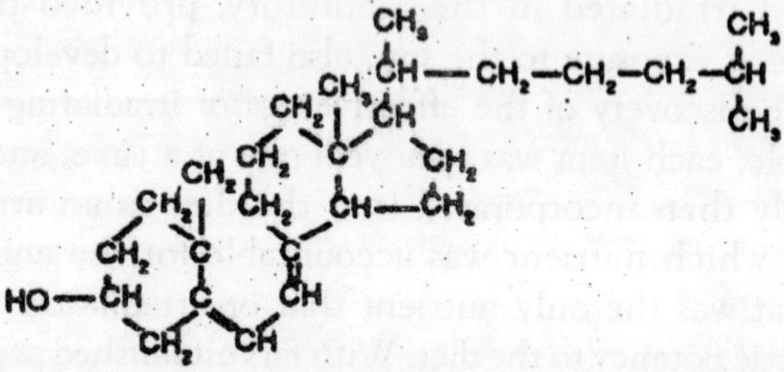

Vitamin D_3

This occurs with vitamin A in fish liver oils and animal fats, but has no concerning with carotene. Chemically it is rather dissimilar from vitamin A and belongs to a complex group of compounds known as sterols.

Chqlecalciferol

In reality many of these have vitamin D activity. The commonest is that occurring naturally in fish liver oils. This is vitamin D_3. Another is calciferol or vitamin D_2 which is made chemically but the same action for man and mammals as D_3. Birds, however, utilize D_2 less well (Harris, 1951).

The term as D_1 has been dropped. It was utilised originally for a compound incorrectly supposed to be the pure vitamin but shown to include D_2 associated with an inactive component.

Vitamin D (Ergocalciferol, Cholecalciferol)

It has been said facetiously that civilisation changed vitamin D from a harmone into vitamin. Certainly the advantageous effect of sunlight is the prevention of rickets was detected long privious there was any indication that vitamins existed. Besides there is hither to some question as to whether a dietary way of vitamin is necessary if the body is regularly exposed to ultraviolet irradiations.

In the mid 1920's Steenbock (1924) demonstrated that irradeation of foods yielded antirachitic activity. Later the phenomenas was shown to be since of activation of sterols in foods. The Steenbock irradiation process was to assist the winconsin Aluuri research and the rayalties obtained assist to start and continue Foundation at the university of winconsin.

The factor accountable for the preservation of rickets was named vitamin D and was consplicuously distinguished from vitamin A also present in oils when it was invented that antirichatic activity remained even after vitamin A activity has been destroyed by oxidation.

Varied types of the vitamin D exist. All arise from irradiation of sterols which are provitamins of precursors of the vitamin. Ergosterol, which occurs in plants, is activated by sunlight to form ergocalciferol (vitamin D_2, calciferol). The form occurring naturally in animal cells is cholecalciferol (vitamin D_3) which is extracted from 7-hydroxycholesterol. The resemplar of the vitamin structure to cholesterol is evident from the more than formula.

Although ultravoilet irradiation produces the vitamin from their sterol precursors, the active forms, ergocalciferol and cholecalciferol are unstable to irradiation. Excessive irradiation will field some toxic products. The vitamin D are not affected by oxidation or by temperatures below 140°c. In acid media the vitamins are comparatively comparatively, the instability increasing with enhancement in temperature. In alkalive solution vitamin D compounds are stable even at elevated temperatures.

The biochemical degradation of cholesterol in the body leads to a host of physiologically significant compounds.

7-hydroxycholesterol available in skin undergoes a complicated succession of photochemical reactions when exposed to sunlight. One of the products is the ring cleaved calciferol or vitamin D_3, which supports to avoid rickets by regulating calcium metabolism.

Vitamin D is a group of fat soluble vitamins. Chemically associaated with the steroids. They are yielded from provitamin, 5, 7-unsaturated sterols, by u.v irradiation. In this transformation, ring B of the steroid is opened between C-9 and C-10, forming precalciferol, which works as a precursor of the vitamin D group. Potent members are Vitamin D_2 (ergocalciferol), extracted from ergosterol, and Vitamin D_3 (cholecalciferol), derived from 7-dehydrocholesterol. Lumisterol and tachysterol are by-products of the synthesis Vitamin D_3 is changed enzymatically in the liver and kidneys to 25-hydroxycholecalciferol, followed by hydroxylation to the highly active la, 25 hydroxycholecalciferol (la, 25- dihydroxy-vitamin D_3); this compound represents the active form of the vitamin in the human, and its behavior similarities that of a hormone, rater than a biocatalyst.

Vitamin D_1 is a molecular compound of lumisterol and ergocalciferol.

Vitamin D_4 is 22-dihydroergocalciferol, yielded from 22-dihyderoergostrol by UV-irradiation.

Vitamin D_3 is present in cod liver oil in especiolly large quantities, and it is also formed in human skin from 7-dehydrocholesterol by the action of sunlight

The vitamin D-Complex is also present in egg herrings, egg yalk, butter, cheese, intak vig liver and edible fungi.

Discovery of the Existence and Nature of Antirachitic Factors

Mellanby found experiment with puppies that their skeletal growth was influenced by some fat-soluble materials or substances in the food. He especially emphaized the prevention of rickets by virtue of an antirachitic property of specific fats.

McCollum and coworkers all but at the sametime found this true also of rats. Morever, cod-liver oil, even after destruction of its vitamin A, still has a particular favorable ascendancy in the prevention of rickets and the promotion of good growth of the bones and teeth. The materials thus found out immediately came to be known as, interchangeably, *antirachitic vitamin* and *vitamin* D. More strictly we say *the vitamins* D, since it has been found that above one materials has the property of preventing rickets.

As more completely explained by Bills (1939) the term vitamin D (collective singular) is customarily applied to any materials possessing the antirachitic property, which is true of at least ten sterols in some degree, but to an important degree of any two of them: named such, vitamin D_2 and D_3.

Each of these can be yielded by the action of ultraviolet light upon its precursor sterol: vitamin D_2 by hence 'irradiating (activating) ergosterol; and vitamin D_3 by hence activating 7-dehydrocholesterol. This 7-dehydrocholesterol has been exhibited to occur in the skin and is the chief "animal" precursor of vitamin D_3, just as ergosterol is the main "plant" precursor of vitamin D_2.

Production of Vitamin D by Irradiation

Hess and independently Steen-bock dislovered it indepen dently that various foods can be endowed with (or enriched in) antirachitic value by irradiation with specific wavelengths of ultraviolet light.

Ergosterol, concentrated industrially from yeast, was found to be a good beginning material for similar activation by irradiation. The particular vitamin D formed by activation of ergosterol is the one known as vitamin D_2. Its commercial forms include calciferol and viosteral. It occurs naturally in fish liver oils, where a blending of vitamins D_2 and D_3 is present, with the former predominating in some varieties, like tuna, and the latter predominating in cod liver oil.

Vitamin D_3 formed by corresponding activation of 7-dehy lrocholesterol, is in some cases more effective nutritionally than vitamin D_2, and it has been detected with the chief natural vitamin D of cod liver oil (and presumably also of milk and eggs). Further activation studies of Bunker, Harris, and Mosher (1940) have also strengthened the authority that 7-dehydrocholesterol is the materials which is activated, and vitamin D_3 the form which is produced, upon irradiation of the skin.

There is a difference of expert opinion whether vitamin D_3 is preferable to vitamin D_2 in the difficult of the more prevention of rickets in infants. In the broader view of child nutrition the reality that vitamin D as obtained in milk and eggs, and in or from fish liver oils, is accompanied by vitamin A is a an important practical cause for preferring similar natural sources.

Chicks and rats differ comprehensively from each other in their responses to vitamin D_2: for the rat it is a comparatively effective antirachitic, and the potency of its preparation is measured by tests with rats. International and U.S.P. units are therefore rat units. "Rat unit for rat unit," vitamin D_3 is more effective in the nutrition of the chick than is vitamin D_2.

Bills stressess the reality that from a given provitamin only one form of vitamin D is yielded by irradiation whatever the wavelength or other conditions. The conditions ascendancy the by-products but the vitamin is forever the same for the same precursor. He adds that processes in which election bombardment is substituted for ultraviolet irradiation produce the same kind of vitamin.

The Prevention of Rickets

Cosistent wisk to Park's definition that has now been

usually recognised, rickets is a condition where in the mineral metabolism is disturbed in similar a source that clarification of the growing bones does not occur commonly. We must keep in mind while reading the literature on rickets that the term has been provided a narrower definition than Park gives it, older definitions turning more upon the histology of the bone tissue while Park's is expressed regarding calcificaion in the developing bone-an inevitably chemical phenomenon.

The failure of normal calcification in the young bone is not usually to be attributed to any initial fault of the bone itself for Shipley found that the rachitic bones placed in normal blood serum started at once to calcify normally; and Kramer and Howland found a subnormal concentration of either calcium or phosphorus (or both) in the blood serum. Thus there are dissimilar sorts of rickets that may be divided in conformity with the nature of the chemical deficiency in the blood.

When the calcium content of the serum is normal but its phosphorus (phosphate ion) content is subnormal then so-called low-phosphorus rickets occurs which corresponds most closely and entirely to the classical histological explanations of the disease. In similar cases not only is a characteristic pathological histology but also a tendency to overgrowth of the cartilaginous or osteoid tissue at the ends of the long bones and at the rib junctions. Since the retarded calcification prevents the bones from receiving normal rigidity at a normal rate, the pressure of the body weight tends further to enlarge the ends of the bones. Together these abnormalities make a typical (and formerly all too familiar) picture recognized in part by the obvious clinical signs of enlarged joints or row of beadlike swellings at the rib junctions and in part by means of the Roentgen ray photograph and the determination of inorganic phosphate in the blood serum; and confirmable at autopsy by histological examination. There has sometimes been a tendency confine the term rickets or "true rickets" to his low-phosphorus kind of.

When the phosphorus content of the serum is normal but its calcium content is subnormal it consequences in a similar gross abnormality but a somewhat dissimilar histology of the bones. This has been known as a "rickets-like condition" or a "second kind of rickets," and is now called as *low-calcium rickets.* Low calcium rickets is usually accompanied by tetany; and it seems impossible that in similar cases the trouble with the nerves as also with the bones is attributable (at least in part) to the deficiency (subnormal concentration) of calcium in the blood.

When both calcium and phosphorus are reduced below normal concentration in the blood serum, calcification is retarded, but the structural abnormality of the bone differs from that of the two types evaluated above. The condition forms *a third kind of rickets* under Park's definition, but due to the difference in clinical and histological appearance it is called *an osteoporosis.*

Unless the mineral supply is deficient, any of the three kinds of rickets can be prevented by vitamin D, which acts to restore to normal the concentrations of calcium and phosphate ions in the blood.

Whence does the vitamin "mobilize" the calcium or phosphate or both? The chief answer appears to be by increasing the *net* absorption of calcium and phosphorous; more being absorbed or less excreted, or both.

This, however may not be quit the comlete story of the action of vitamin D in the prevention of rickets.

Sometimes the calcium or phosphorus mobilized through the blood stream and made available to the developing ends of the bones starts to come from stores already in the tissues.

There are also connotations that a local factor, affecting, as Hess expressed it, the "anchorage" of the available calcium and phosphate ions in the end of the growing bone, may play a part in certain cases. Morever, Schneider and Steenbock have offered authority of particular influence of the vitamin upon phosphorus metabolism like to favor the deposition of calcium phosphate as bone mineral in the cartilage of the developing bone ends, at the expense of the soft tissues, hence inhibiting the common development of the body.

In this connection the report of Zetterstrom that phosphorylated vitamin D activates alkaline phosphatases from intestine, bone, and kidney, hence functioning in the absorption deposition and retention of phosphate, is of interest.

While scientific analysis is still not complete, the difficult of prevention of rickets is now adequately solved that few new cases of severe rickets are noticed and the mild rickets now-a-days occurring may in large part be considered as incidental to the fluctuating fortunes with which the soft tissues and the developing bones complete for the phosphate which the blood stream brings.

The prohibition of rickets by the exposure of the body to sunshine or ultraviolet irradiation comes about through the production of animal vitamin D" (vitamin D_3) in the skin. Notice, too the efficiency of the "mechanism" which at the same time fastens the circulation of blood through the skin, bringing precursor and taking active vitamin into body.

The Promotion of Growth and Development

When severe rickets was more common, its outward signs of bowlegs and knock-knees appeared more explicit in so far as common growth means increase of soft tissue; and the more active the soft tissue growth the less the supply of phosphorus for the deposition of bone salt which is essential to the normal hardening of the growing bones. If the bones are prevented from abundant hardening and also are made bear an increasing body weight, the bending of the bones and enlargement of the joints will be worse, the greater the weight put them. Under similar conditions, the greater the growth, the worse the rickets." But, with the discovery and common use of vitamin D, the situation has greatly changed.

The stress at present is on the liberal nutritional intake of calcium, of phosphorus, and of vitamin D. When the intakes of calcium and phosphorus are abundant to the needs of the development and growth of both soft tissues and bone, liberality of vitamin D tends to eveourage growth; more specially liner growth, i.e., increase in the length of the skeleton, perhaps particularly of the legs.

The superior height and erectness of the carriage which in some countries has so far been considered "a characteristic of the upper class" is following long straight legs and is now being largely conferred upon the children of all economic conditions through the discovery of vitamin D and the widespread utilize of fish live oils.

Like some other of the constructive developments of the newer knowledge of nutrition, the ascendancy of liberal intakes of vitamin D upon growth, when intakes of calcium and phosphorus are also liberal, was first displayed by experimentation with laboratory animals of short natural life cycle and then confirmed and extended by clinical experience.

Apart from to its role in the prevention and cure of rickets, vitamin D was found (Sherman and Stiebeling. 1929,1930) to excourage development and to facilitate the normal calcification in growing bone even when the mineral metabolism is well more than the rickets level.

Along with other factors, vitamin D is of significance in the formation of normal teeth and protection against dental caries.

There is clinical evidence that children who have been protected from rickets cod liver oil are less susceptible to respiratory disease than those who have had rickets , but it remains an open question in how far the stronger respiratory system is due to vitamin D and in how far to the vitamin A which the cod liver oil furnished.

The repeated emphasis which is of greater significance laid by Jeans and Stearns upon the reality that while vitamin D is in this regard perhaps even more hopeful to physical growth than heretofore percepted, ons's confidence in the vitamin must never be permitted to make one less carefully attentive to the importance of liberal calcium intake. Increased calcification means increased retention of calcium in the body, and the body must receive large amounts so as to retain large amounts.

The vitamin must never be expected to decrease the calcium requirement. Rather, it shpports the body to make optimal utilize of a generous calcium intake. The intake of calcium must forever be generous is optimal calcification and devolopment of bone are to be secured.

The Properties of Vitamin D

The sterols which are significant ways of vitamin D are known a pro-vitamins. There are many of them, but only two of importance, 7-dehyrocholesterol and ergosterol. The provitamin in higher animals and invertebrates is 7-dehydrocholesterol; ergosterol is present in the plant kingdom, in yeasts, and particular tye of molds.

The pure vitamins D are white, crystalline, odorless substances that are soluble in fats and fat solvents like ether, chloroform, acetone, and alochol. They are resistant to heat, oxidation, acid, and alkali. The vitamins are consissted ofcarbon, hydrogen, and oxygen combined in chemical structures that are much alike. Crystals of vitamin D_2 and its chemical formula are shown in Fig. 14.5. The relative antiricketic potency of vitamin D_2 and D_3 seem to be no dissimilar for the human being; consequently, in present discussion the term vitamin D *will be* used.

The Unit of Measurement of Vitamin D

Like that of vitamin A, Vitamin D potency is expressed in International and USP units. (Other vitamins are measured in units of weight). The amount of antiricketic activity brought about by 0.025 ug of the International Standard, pure crystalline vitamin D_3 is an International Unit. The unit was embracad by the World Health Organization in 1949 and has been used ever since. The USP unit is the same as the International Unit.

Wih a view to measure the amount of antiricketic activity, rats are usually employed as the experimental animal. Under standardized conditions of feeding and care, and by using standardized procedures for conditioning bone change, the potency of vitamin D regarding International Units is determined.

Measurement And Expression Of Vitamin D Values

Vitamin D values are often measured either by means of the "line test" or by determination of calcium or total ash in a representative bone.

McCollum and his coworkers have found in their autopsies rats which had been on deficient diets, that a rickets-producing diet of a sufficiently drastic character may consequence in condition wherever the cartilage and adjacent portions of the metaphyses of the long bones are completely free from visible deposits of calcium phosphate. If, then, similar rats are provided an effective antirachitic treatment of any kind, then with a few days a line of freshly laid down deposits of calcium phosphate which by appropriate staining can be made prominent to histological examination. This is the *line test.* Bills, Honeywell, et al. (1928,1931) elaborated the technique and gave a very full description of it utilised by them. It was also adopted by Steenbock and his coworkers for the routine testing concerned to with their control of the manufacture of irradiated foods and drugs under the Steenbock patents.

The line test has been followed as the basis for *the United States Pharmacopoeia* unit for vitamin D potency. The existing procedure provides for the comparison of the degree of healing caused by the test material with that caused by a known quantity of a standard reference material in a corresponding group of animal under identical conditions. In this source the impression of variations in the susceptibility to rickets of different groups of rats, due to environmental and other factors, are subsidined.

The International standard reference material has been a crystal line preparation of vitamin D_2; and the International unit has been explained as the antirachitic activity of 0.25 microgram of this standard material. The Expert Committee on Biological Standardization of the World Health Organization offers a crystalline preparation of vitamin D_3, the unit of activity in this case also being defined as that of 0.025 microgram of the reference material. The *United States Pharmacopeia* has followed this new International standard.

To develop rickets in rats for the line-test method of assay, greatly unbalanced mineral intake are used, the usual rachitogenic diet containing four to five times as much as calcium as phosphorus, that is a very much wider ratio than is likely to be encountered in human experience.

Stiebeling has utilised a dissimilar method wherein the basal diet fed the experimental animals (rats) is sufficient in its mineral content as also in all organic factors other than vitamin D, and where in the vitamin D value of the material under investigation is judged by the relative amount essential to allow a standard degree of calcification under fixed experimental conditions.

Young rats receiving the basal diet only (and on irradiation) serves "negative controls"; those receiving the basal diet plus substantial vitamin D (or irradiation) serve as "positive controls." others receive the basal diet plus graded allowances of the food or other material which is being examined for vitamin D value.

At the final opinion of a four or five week experimental period the degree of calcification is found by conditioning the percentage of ash or of calcium in the freshly removed femur. Within specific limits, this is found to be dependent upon the relative amount of vitamin D supplied. The benefit in calcification shown by the test animals over the negative controls must be due to of the material under test given the experiments are properly conducted and in enough numbers to avoid vitiation of consequences through individual variation. It is suggested that the "end point" for comparison be one half as much improvement in calcification over negative controls as is yielded by supplying an abundance of vitamin D (positive controls). The amount of the food needs to "reach this end point " (maintain the standard rate of calcification) may be considered as inversely proportional to its richness in vitamin D.

Transformation of Provitamin D to the Vitamin

The transformation of provitamin D to the vitamin by irradiation is accomplished by a modify in the structure of the compound; the second ring in the provitamin D structure opens (note ring B, Fig 2). The conversion process has been studied in the laboratory with the finding that a succession of intermediary products are formed privious vitamin D is attained; further irradiation beyond the vitamin D stage produces toxic products lacking antiricketic potency.

The Human Requirement of Vitamin D

Jeans and Stearns hold the view that for many persons some vitamin D in addition to that generally obtained by exposure to sunshine is essential for optimal utilize of calcium and phosphorus in the body. This may be obtained by special irradiation of the skin, by utilize of similar foods as contain vitamin D, by the taking of the vitamin in some more concentrated form, or in a combination of these ways.

Jeans and Stearns indicate that we should recognize three phase; of the vitamin D need: (1) that for normal development and mineralization of the skeleton comprising the teeth, of infants and children; (2) that for maintenance of these skeletal structures in sound condition throughout adult life; and (3) that to supply the special needs of pregnancy and lactation.

Still the is no an agreed scientific opinion as to what forms a normal or optimal rate of mineralization of the developing skeleton. Jeans and Stearns state that this is most quick in early life, and that among the estimates of desirable calcium retention "the higher figures are approach to be a safer

During infancy the vitamin D requirement relies to some extent upon the rate of growth. The more quickly the infant grows the more significant it is to safeguard him against rickets:

also the liberal allowance of vitamin D, gives as similar a safeguard, itself tends to increase the rate of growth. Infants whose daily food includes a quart of milk containing 135 units of vitamin D appear to be entirely protected from rickets; but additional vitamin D up to a total of 300 to 400 units seems to support a rate of development and growth "nearer the top of the normal range."

The observations of Jeans and Stearns and of Macy and coworkers indicate that increasing allowances up to an undetermined level between 135 and 300 units of vitamin D per day tends toward a more uniform ability among infants to assimilate calcium from the food. This seems to be true even of infants fed human milk.

During Childhood there is requirement of continued emphasis upon liberal intakes of calcium and of vitamin D if optimal growth of bones and teeth is to be maintained. Jeans and Stearns (1939) said the requirements of children in this respect do not appear to have been as completely realized as the needs of infants, most likely because it is less easy to make the child's requirement obvious. Beyond emphasizing the view that there should be no let-down of calcium or vitamin D intake as infancy develops into childhood, little can yet be said as to the child's quantitative requirement.

Jeans and Stearns have studied fifty children aged 1 to 12 years with varying intakes of milk and with and in absence of added vitamin D in the form of cod liver oil to the extent of 300 to 400 added units of vitamin D per day. They found that children different in their need. If a child used calcium efficiently in absence of the added vitamin D, the extra of this to the diet had little effect; while if the original utilization was poor, the ingestion of additional vitamin D increased the retention of calcium.

Whether adults require vitamin D and if so to what extent axe questions upon which there is little evidence beyond the fact, so carefully an repeatedly stressed by Jeans and Stearns, that vitamin D does not lower the minimum requare for ingested calcium.

In pregnancy and lactation it is undoubtedly wise to be liberal with the vitamin D allowance.

The National Research Council's Recommended Allowances give 400 units per day for each child or pregnant or nursing woman.

Sources of Vitamin D

The antirachitic values of fish liver oils, of irradiated ergosterol, and of ultraviolet irradiation of the body, are so great that they have absorbed attention to most likely a greater extent than is desirable. The availability of significant amounts of vitamin D in egg yolk, all milk, and butterfat has, however, been established. That in some cases the authority of its presence has failed to seem may be attributed to, in large part, to the utilize in diverse experiments of so drastic a rickets-producing diet that symptoms of rickets developed in the experimental animals despite the presence of amounts of vitamin D (in the food fed for test) which would have been measurable under more nearly normal conditions. Mellanby, working with puppies and using basal diets not so drastically deficient as are the rickets-producing diets utilized by most of those who work with rats, had no difficulty in showing the vitamin D value of green vegetables and of butterfat. Shipley, Kinney, and McCollum (1924) exhibited the presence of vitamin D in ether extracts of green leaves. Also in the work of Sherman and Steibeling which demonstrated a considerable vitamin D value in milk, the diets were so arranged and controlled that the influence of the milk in developing deposition of calcium phosphate in the growing bone must have been due to the vitamin readily than the calcium or phosphorus factor of the food, since the amounts and ratios of calcium and phosphorus in the intake were strictly controlled.

As is true of the vitamin A, so also for vitamin D there appears to be survival value for the species in the capacities that have been developed by the maternal organism to accumulate the vitamin and pass it on to its young through the fat of the egg yolk or of the milk as the case may be. The concentration is higher in the egg fat, but the total amount of vitamin D obtained through the milk fat is considerable due to the consumption of the latter not only in the shape of milk itself but also of cheese, butter, cream, and ice cream. Vitamin D is sufficiently stable, except under irradiation, so that no thought full loss requare be feared in the formulating or storing of similar foods as these, or in ordinary cooking operations. Direct experimentation has displayed it stability at autoclave temperatures.

Vitamin D milk is the term commonly utilised for market milk of enhanced vitamin D value.

Researhers differ as to the praticability of materially increasing the vitamin D value of milk by the irradiation of the cow.

Three methods of enrichment of milk in vitamin D have been on practical use: (1) the addition of a concentrate of natural vitamin D; (2) feeding irradiated material to the milch cow; (3) irradiation of the milk itself.

The choice between these three methods depends partly on economic factors, and partly upon considerations having to do with the difference in kinds and relative proportions of the vitamin D or vitamin D introduced by the respective methods.

Whatever may be the the final agreed view as to the relative merits of these dissimilar kinds of enrichment, it is already established that vitamin D milk is a highly effective means of administering vitamin D.

With milk of similar greatly enriched vitamin D value now usually available, and with the wide common distribution of cod liver oil, of the more potent liver oil of the halibut, and of the still more potent *percomorph liver oil*, it is appropriate enough to place main dependence upon these highly potent and now well standardized sources. But the natural vitamin D values of foods are scientifically significant to our understanding of the process of nature and how children got along even as well as they did privious vitamin D was foundout.

Storage of Vitamin D in the Body and its Transfer from Mother to Young

Both carefully controlled laboratory experiments with rats and clinical experience with mothers and infants havè dissplayed that the antirachitic vitamin can be stored in the body, and to a very greater extent. The nutritional condition of the mother influences the storage of vitamin D in the body of the baby privious its birth, and her bodily store of this vitamin (or lack of it) affects the antirachitic potency of her milk.

Vitamin D taken by a pregnant or nursing mother is therefore, a nutritional asset both to herself and to the child.

Hess and co-workers encountered cases in which young rats which they had intended using as test animals proved "refractory to rickets," following of the highly antirachitic character of the diet of mother and the reality that the litters were small. This, of course, is a converse demonstration of the clinical experience that severe rickets is more repeated among the children of the poor whose mothers are living on food of inferior mineral and vitamin content, often in dark tenements, and also have borne and suckled many children.

Functions of Vitamin D and the Effect of a Deficiency

Bone Formation. One of the chief function of vitamin D is the maintenance of the normal processes of bone formation. A deficiency of the vitamin in the growing young consequences in rickets. The disease will develop in infants even though the diet provides an ample intake of calcium and phosphorus unless vitamin D is given in the diet or well exposure to sunlight is feasible. Late rickets (osteomalacia), a condition corresponding to rickets in children, occurs in adults.

In rickets the bones are susceptible to deformity when exposed to ordinary stresses. A general and severe modify in the skeletal structure is enlargement of the ends of the long bones in the legs and arms. If mineral is not deposited in the growing bone cells, the weight of the body reasons the noncalcified portion to flatten and mushroom outward, offering the appearance of enlarged joints. Other bone changes comprisse the *ricketic rosary,* a row of beadlike protuberances on each side of the chest at the function of the rib bones and joining (costal) cartilage (Fig. 3), bowed legs, and knocked knees—all these due to failure of the bones to calcify normally.

The method of action of vitamin D in terms of bone calcification remains comparatively obscure. Intramuscularly injected radioactive calcium was deposited in the bones of rickets chicks with no vitamin D in the diet. This indicated to the investigators that the chief function of vitamin D might be that of facilitating absorption of calcium from the intestine since calcium deposition in bones could proceed in absence of it (9).

Calcium Absorption and Metabolism : That vitamin D exerts a favourable ascendancy on the absorption of calcium is witnessed in studies with animals and human beings. Figure 4. illustrates the decrease in calcium excretion in children with rickets after the administration of vitamin D; the fecal exertion is markedly reduced. With children 1 to 12 years of age, Jeans and Stearns found that some used calcium efficiently in absence of added vitamin D; others needed a supplement. Stearns pointed out that there is no source of identifying from outward appearances which children requare the additional vitamin, which do not, thus recommended that vitamin D be provided to all children. Numerous balance studies with normal adults show no improvement in the utilization of calcium with additions of ways of vitamin D except in the period of pregnancy and lactation.

The absorption of phosphorus appears not to be influenced by vitamin D. For example, vitamin D-deficient animals absorb phosphorus equally well, with and in absence of an additional supplement of the vitamin.

Growth. Rickets reasons a slowing of growth. Animals wherein experimental rickets has been induced grow more gradually than their normal controls with an sufficient vitamin

D intake. Infants also fail to grow normally in absence of a suibstantial amount of the vitamin. Infants provided 340 to 400 IU daily of vitamin D were found to grow more in length than those receiving 60 to 135 IU. Giving above 600 IU daily did not increase the development rate further; increasing the vitamin D intake beyond 1800 IU daily decreased growth.

Incidence of Vitamin D Deficiency

Detectable rickets ollinically is not general in the United States today. The widespread preventive (prophylactic) utilize of ways of vitamin D has all but eradicated this once general scourge of infancy. Fortification of all evaporated milk and much of the fresh milk with vitamin D all but controls rickets automatically in the United States. Osteomalacia, too, is rare in this country. In conformity with a report it is general among women in the Orient and parts of the Near East in the period of pregnancy and lactation, increasing in severity with each reproductive cycle.

Hypervitaminosis D

Big dosages of vitamin D may have toxic impression if continued for a period of time. Giving only two massive doses a year is utilised by some physicians as an adopted means of providing vitamin D. The level at which deleterious impression result continued high dosages varies with individuals. Some indicates symptoms with an intake of 50,000 IU per day for a few weeks, others tolerate as much as ten times that amount for longer periods of time. An intake of 1800 IU daily may be mildly toxic in children if continued over a period of time. The early symptoms include loss of appetite, thirst, lassitued, followed by nausea, vomiting, diarrhea, abdominal discomfort, and loss of weight. The blood calcium chronic condition that has been available for some time becoming acute only when a high degree of toxicity is reached.

Vitamin D preparations on the market are of similar potency that dosages are expressed in drops, formulating it convenient to give above is wanted. In commonly used preparations around 50 drops measure one-half teaspoon with a potency of 3000 IU. Most multiple vitamin capsules give 500 IU or more per capsule.

Suggested Permission for Vitamin D

The Recommended Allowance for vitamin D for late adolescence is 400 IU daily, the same as is provided as per recommendation throughout childhood. Normal adults seem to require little, if any, supplemental vitamin D. It is usually believed, nevertheless, that for adults with little exposure to the direct sunlight some vitamin D in the diet may be advisable, notwish slanding the lack of direct evidence to substantiate the presumption. During pregnancy and lactation, when the calcium requare is increased, the Recommended Allowance is 400 IU daily. The requirements of infants and children, and further discussion of the requirementss during pregnancy and lactation seem in those respective later chapters.

Vitamin D : Its Sources

Vitamin D is not comprehensively distributed in nature. The best ways are fish oils, particularly liver oils, the concentration of which varies with the species. Milk, butter, and egg yolk are the only foods in the ordinary diet that contain vitamin D, and in these the amount is small and variable. Vitamin D present in food is stable in ordinary cooking, heat-processing treatments, and storage. Milk is selected for enrichment with vitamin D since it is a rich way of calcium and phosphorus, the minerals that requare vitamin D for absorption and use. It is consumed by the young, who especially requare the vitamin for skeletal development.

There are fish liver oils on the market and concentrates of irradiated ergosterol; also vitamin D is involved in diverse multiple-vitamin combinations. Some of the preparations of vitamin D are in oily solution, some are aqueous preparations. The potency as connoted on the label is assured of accuracy by the reliability of the pharmaceutical house manufacturing the product and the surveillance of the Federal Food and Drug Administration over the authenticity labels.

4

Vitamin A : Its Nutrition

Vitamin A is stored in the liver and during adult life the daily requires can be supplied to a considerable extent from these reserves. Volunteers who intentionally consumed diets very low in vitamin A were skilful to continue doing so for several months in absence of developing any gross signs of deficiency. But despite this there is common agreement that it is better to consume sufficient to guard against any chances of deficiency and, in the table, the recommended levels of vitamin A vary very little.

Whether taken as vitamin A itself or in the plant form carotene, this factor is forever changed into vitamin A used as similar in the animal body. But the two forms are not of equal physiological value, unit for unit. It is usually reckoned that the number of international units needed in the form of carotene is about twice the number required as vitamin A itself. Even this ratio is not the same for all foods. The carotene of green vegetables supplies to the body about half as many units of vitamin A as the number obtained by chemical estimation, but yellow vegetables like carrots supply only about one-fifth or one-quarter (Graves, 1942).

Historical Backdrop

Vitamin A was the first to here lognised as a component. In 1913 Davis and Mc Collum of particular food invented that rats could not become skilful to develop as diet consisting of protein carbohydrates and salts.

In two laboratories experiments were in progress that led to the discovery of vitamin A in 1913. McCollum and Davis at the University of Wisconsin fed young rats a purified diet considerd to be enough at that time. It contained protein, carbohydrate, mineral salts, and fat (lard), whole of the nutrients then recognized. They used milk sugar in this experiment, which added sufficient impurities of nutritional value for the young to grow fairly well for a while. However, after a time the animals ceased growing. Adjustments were made in the relative amounts of the nutrients in the diet in an attempt to prohibit the cessation of development but to no avail. When butterfat or egg yolk was added to the diet, development was resumed. The addition of olive oil, on the other hand, had no effect. It was explicit that there was a difference in the nutritive value of olive oil and lard, and butterfat and egg yolk. At the Connecticut Experiment Station in the same year, Osborne and Mendel fed Young rats a purified diet comprising protein, lard, starch, and protein-free, fat-free milk that given the minerals, some lactose, and other water-soluble substances. As in the Wisconsin experiments, after some time growth failed. With replacement of lard by butter, the rats grew and matured normally. Eggs yolk and cod liver oil in the diet gave the same response as butter; these fats given the *some what* that was lacking in the purified diet.

McCollum, and Osborne and Mendel, concluded that the necessay nutrient missing in the purified diets was soluble in fat. Knowing the solubility of the materials helped to speed the discovery of vitamin A; different compounds were hence excluded from consideration. Further assist came from Steenbock's studies (1919) at the University of Wisconsin in which root vegetables were fed as a way of the-soluble vitamin.

Yellow sweet potatoes and carrots assist normal development and even supplied sufficient of the unknown materials for reproduction. Rutabagas, red beets, parsnips, potatoes, and sugar beets failed entirely to give the required nutrient. Steenbock tested many plant materials, among them yellow corn and white corn. The animals fed yellow corn as a supplement to the purified ration grew normally and were skilful to reproduce; animals fed white corn failed to grow and developed deficiency symptoms. It is of great interest that McCollum, in 1908, heard Wisconsin farmers who were attending a meeting at the University maintain that yellow corn was better feed for hogs than white corn; that was eleven years privious Steenbock's studies confirming their contention. Thus more over solubility in fat, another characteristic of the nutrient was recognized, the association with yellow color. It had been demonstrated that animals were not able to synthesize the yellow pigment carotene. This gave a lead that it must be given in the diet. The question still remained, however, as to the relationship between the yellow plant substance and the animal fat substances, both of which supplied the fat-souble vitamin. After ten years later, in 1929, crystalline carotene was exhibited to have vitamin A value by a German scientist Hans von Euler-Chelpin.

Vitamin A and Carotene : Its Properties

Vitamin A is an organic compound found only in the animal Kingdom. It is believed to be present in whole species of fish, birds, and mammals. Yellow plant pigments of the carotenoid group *alpha-*, *beta-*, and *gamma-* carotene, and cryptoxanthin are precursor of vitamin A. The body posssess capacity of converting these carotenoid compounds contained in the diet into vitamin A. Indeed the origin of whole vitamin A in animal products comes directly or indirectly from the carotenoids of plants. Fish consume marine plankton; herbivorous land animals eat vegetable foods; and the carnivorous animals reliy indirectly on herbivorous sources.

Two types of vitamin A have been recognised: vitamin A_1 which occurs in the livers of salt-water fish, and vitamin A_2

found in the livers of fresh water fish. Supposing that the two vitamins differ minutely in chemical structure they are alike in physiological functions. Vitamin A_2 contains an extra double bond in its ring structure. In the present discussion they will be treated as one, only the term vitamin A will be utilised.

Vitamin A is a pale yellow materials composed of carbon, hydrogen, and oxygen. The chemical formula for vitamin A_1 is provided in. It is soluble in fat, not soluble in water, stable to heat, that is, to ordinary cooking temperatures, and destroyed by both oxidation and ultraviolet irradiation.

The pigments of carotenoid are compound are the consequence of combination of carbon and hydrogen. The crystals of the carotenoid pigments are a deep red color but in solution intensely yellow. Their properties of solubility and stability are such as those of vitamin A. Beta-carotene is more ably converted to vitamin A than alpha- or gamma-carotene, or cryptoxanthin. The chemical structure of beta-carotene is similar that it oxidizes to form two molecules of vitamin A; the other provitamins form only one. However, in view of of clarity, the carotenoid groups will be spoken of as *carotene,* as though it were one substance.

Chemical Structures and Interrelationships

Vitamin A is a colorless, fat-soluble substance. It occurs in relative abundance in the fats of milk, eggs, livers, and fish. It is found in high concentration in fish liver oils. It now seems that there are two correlated substances, vitamin A_1 (which predominates in mammals and salt water fish) and vitamin A_2(which predominates in fresh water fish liver oils). It is easy and scientifically admissible to continue to utilize the term vitamin A as a collective singular to cover these two substances functioning inevitably as if they were one.

At least four substances (alpha-, beta-, and gamma-carotene, and cryptoxanthin) are known to be precursors of vitamin A, and are easily changed to vitamin A in the body. Hence the presence of any of these precursors in a food adds to its vitamin A *value*, though not to its vitamin A *content*. The fish liver oils contain vitamin A and practically none of its precursors. Foods generally however, owe their vitamin A values at least partly to the precursors which they consist of.

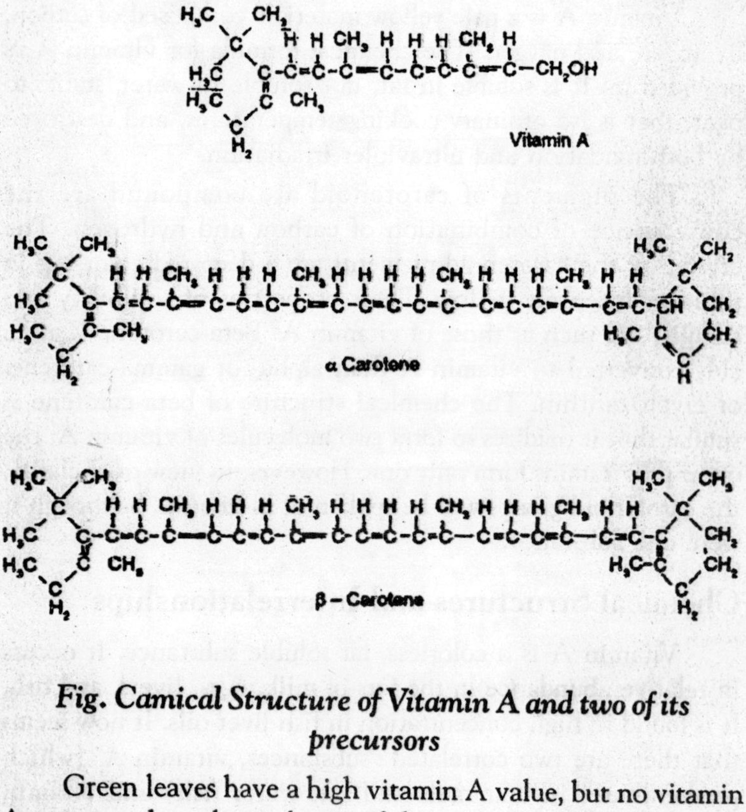

Fig. *Camical Structure of Vitamin A and two of its precursors*

Green leaves have a high vitamin A value, but no vitamin A content; the value is due to of the presence of the precursors. This is reality with the vitamin A values of all plant orangs. Animal organs contain both vitamin A and some unchanged precursor, and this is true of milk and eggs also.

The adopted structural formulae for vitamin A and its two best known precursors are displayed in Figure 4.

It will be seen that the structural formula of the beta-carotene molecule, $C_{40}H_{56}$ is symmetrical, with a "double bond" in the centre (Fig 4). By dividing at this point, each half taking on both an H and an OH, it is theoretically capable of prducing two molecules of vitamin A, $C_{20}H_{30}O$; while by parallel reactions the other precursors known would each product only one molecule of vitamin A, corresponding to the half of the precursor molecule here shown on the reader's left, since only this half comprises the vitamin A structure.

Practically the body may get significantly less than the theoretical products of vitamin A from these precursors as they occur in foods Even so, the vitamin A values of similar green leaves as those of kale are high &s shown by feeding tests (Sherman and Todhunter, 1934).

Vitamin A : Its Nutritional Analysis

It is of above historical interest that the existence of vitamin A was found out through experiments (made independently and all but simultaneously in 1913 by McCollum and Davis and by Osborne and Mendel) where in it was found that young animals on diets good in other regards would continue to grow and thrive or would cease growing and sicken, according as the sole fat in their food mixtures was butterfat or lard. Further work exhibited that egg fat and cod liver oil similar butterfat in this respect while most commercial fats and oils are more such as lard. The property, by virtue of which the maternal organism concentrates vitamin A in the material (milk or egg) given for the offspring, seems to have a high *survival value* for both mammalian and avian species.

The significance of vitamin A to development and growth is evaluated in the experiments and observations, just mentioned, through which the existence of this factor was discovered. *Its relations to the eyes* are also of great significance.

Vitamin A is directly included in the chemical changes upon which vision depends, and it is commonly found that a diminished efficiency of adaptation of vision to a changed intensity of light (dysadaptation, hemeralopia, nightblindness) will be among the earliest observable impression of a shortage of vitamin A in any normal person.

The vitamin A which functions in vision is doubtless brought to the retina by the blood, a lowered concentration of this vitamin in the blood may be hoped to accompany or precede the lowering of visual efficiency; but the authority on this point is not completely clear and consistent.

The physical authority obtainable through biomicroscopy. has been developed as a diagnostic method by Kruse (1941).

It displays that vitamin A deficiencies are considerably more prevalent than the chemical methods connote. This clear differences may or may not be satisfactorily explained by the reality that what is found by chemical analysis of the blood is something rapidly responsive to nutritional intake while the Bitot spots and perhaps other physical signs divulged by the biomicroscope are not so quickly changed by a modify of intake. A question still remains as to whether a deficiency has been completely cured as soon as the blood chemistry has been brought to normal.

In conformity with the methods and interpretations of biomicroscopy more vitamin A value should be provided, both for prevention and for complete cure, than has been thought necessary so far.

International Unit of Measurement

In International Units (IU) Vitamin A is measured. The term itself indicates the universality of the unit of measurment. The Health Organization of the League of Nations proposed the global unit for vitamin A to replace the varuiys dissimilar units in utilize at that time. The International Unit of vitamin A activity as adopted by the current official body, the Subcommittee on Fat-Soluble Vitamins of the World Health Organization is equivalent to 0.3 *ug* of vitamin A alcohol. Equal biologic activity is afforded by 0.6 *ug* of the provitamin beta-carotene.

Pharmaceutical companies of the United States utilize the United States Pharmacopeia (USP) Unit to express the vitamin A potency of their products. The USP Unit is exactly the same as the International Unit. The U.S. Pharmacopeia is a book mentioning drugs, chemicals, and medicinal preparations published under the official evidence of the Food and Drug Administration.

Demonstration and Measurement of Vitamin A Values by Feeding Experiments

When a diet lacking vitamin A value but sufficient in all other respects is given to a young growing rat, development continues for a shorter or longer time consistent with to the extent of the stores of this vitamin which the body possesses, after which development ceases and there generally soon sets in a loss of body weight and a condition of decline in health. About when the body's reserves of vitamin A have been exhausted and development ceases, a large proportion of the experimental animals start to develop a characteristic disease of the eye. This generally begins with a swelling of the lids of one or both eyes or with indications that the eye is becoming unduly sensitive to light. Because of this there commonly develops an inflamed and catarrhal condition of the conjunctivae, with a bloody or purulent discharge, the lids becoming scabby or sticky. This, with the swelling of the lids and the sensitiveness to light, sometimes consequences in the eye being found entirety closed. If the eye condition is not treated, and the animal continues to live, the cornea may become affected and blindness results. On the other hand the eye condition if not too far advanced, usually improves quickly upon feeding the animal with any food of enough vitamin A value.

Thus a demonstration of vitamin A value in a food may be made by feeding the food to be tested to a young rat which has recently ceased to grow and begun to develop eye trouble form vitamin A deficiency as just described. Under these curcumstances if the test food has important vitamin A value it will usually induce resumption of development and recovery from the eye disease.

And similar testing of foods can be made quantitative by standardizing the conditions and measuring the rate of development induced by the feeding of graded allowances of the test food and of a reference material of accurately known vitamin A value.

Beta-carotene was prepared by the Health Organization of the League of Nations of which 0.6 microgram had, by definition, one International unit of vitamin A value. The U. S. Pharmacopeia unit of vitamin A value is the same; but the Pharmacopeia organization provides a reference material, more easy for ordinary utilize than carotene, in the form of a cod liver oil whose value has been measured in comparison with the International Standard.

In recent past, the World Health Organization has offered, as a new reference standard, crystalline vitamin A acetate. At present (1951-52) both this and the beta-carotene standard are in use. Description of the method should also comprise definition of the oil in which the carotene is dissolved.

The same test may be made both qualitative and quantitative by permitting growth of the typical sore eye privious the feeding of the test food is sstarted; but if this involves thought full illness in the test animal it is apt to render the consequences more variable, so that for quantitative work the test feeding is generally begun privious the test animal becomes seriously ill, while at least one member of each litter or set of test animals is continued on the basal diet alone as a *negative control* in which the typical growth of the vitamin A deficiency disease can be observed. Quantitative measurements of the relative amounts of vitamin A in food can be made by means of sufficiently numerous and carefully controlled feeding experiments with rats which are sufficiently given with all other nutrients which they require.

The general plan to start with normal young rats 21to29 days old, feeding them a basal diet good in all other respects but free from vitamin A. When any surplus store of this vitamin which the body contained at the beginning has been depleted (or adequately diluted by the development of the animal) the body weight increases and usually remains nearly stationary for a few days. At this moment, if not before, each of the animals is placed in a separate cage and kept under controlled conditions. One (or more) of each litter continues to receive the basal ration only, until it dies from dearth of vitamin A. Thus serving as a negative control. Other animals are fed, in addition to the basal diet, systematically graded amounts of the reference material or of the food which is being tested.

Natural biological variability formulates it possible that the consequences of individual experiments of the sort will vary above should the data of chemical experiments conducted *in vitro*; but when substantial numbers of animals are used and the consequences are averaged, the average weight curves show a fairly continuous gradation in conformity with the amount of vitamin A which the animal actually received.

The accompanying graph shows the consequences received by Batchelder in negative controls and in test animals fed at diverse different levels of vitamin A intake (allowance of test food), each curve being the average consequence obtained with nine test animals. The beginning point of these curves is the end of the *depletion period* and starting of the *test period*.

In the experiments of Todhunter, vitamin A values were conditioned by a method of single feedings (Sherman and Todhunter, 1934). The particular kind of consequences summarized in are of interest in demonstrating the resemblance of weight-curve resposne to equivalent vitamin A values whether in the form of the vitamin itself (as in cod liver oil) or its precursor (as in the case of carotene or kale) or a mixture of the two; and also in displaying the quantitative relationship of weight gain and survival period to the amount of the single feeding of any of these unique ways of vitamin A value. The result of this relationship is that the area, bounded more than by the weight curve resulting from a provided feeding and below by the negative control curve and the base line, is (under amply standardized situations) quantitatively proportional to the vitamin A value of the weighed amount of test material utilised in the single feeding. This formulates it feasible to measure quantitatively the vitamin A valuses of substance which are only rarely available, or of individual samples of fresh foods in absence of subjecting them to the feasibelity of any decline in vitamin A value due to drying or storage.

Vitamin A : Its Importance of Foods

The carotenes which constitute the chief precursors of the vitamin A of normal human nutrition, as well as of ordinary farm animals are formed in plants. The functional interrelationship of the C250 tenes and chlorophylls in the nutritional processes of the plant is still a difficult of active research. The action of light does not seem to be strictly for the formation of carotene, for MacLeod and coworkers have observed an example of increase of vitamin A value in a root (sweetpotato) on storage, and some small increases have been reported in etiolated seedlings. But, as displayed by Drummond and Coward, the pronounced growth of vitamin A value in a seedling accompanies the appearance of green color, so that carotene and chlorophyll formation normally go on in a more or less parallel source in young plants and green leaves. The green leaves of food and forage crops and of pastures serve directly and indirectly to supply us with vitamin A values. The high vitamin A values of typical pasture grasses have been properly measured by Woods and coworkers (1932,1935). When we eat green leaves, we absorb (with varying degrees of completeness) the carotenes which they contain; and our bodies modify these carotenes (again, with different degrees of completeness) into vitamin A which hence becomes available either for immediate nutritional requires or for storage in the body. And the carotenes of a wider range of green leaves are brought indirectly into the service of human nutrition through farm animals—milch cows and laying hens. The cow and the hen render important services to human nutrition by bringing into the easily digestible and assimilable forms of milk and eggs an important share of the nutritive values of leaves of coarser quality than we care to eat or which would exceed the quantity that we easily digest.

In any comprehensive access of the difficult of food supply, the outstandingly significant ways of vitamin A value are the green and yellow vegetables, eggs, milk, and similar products of milk as contain a significant proportion of its fat (e.g., butter, cream, ice cream, and whole-milk or "cream" cheeses).

The vitamin A values reported for green leaf vegetables vary quite widely, even for the same species. This is doubtless attributable in part to actual variability of the substance and in part to laboratory errors following both the utilize of methods still in process of quantitative development and to differences in the aptitude and experience of the laboratory workers.

The high vitamin A value of the completely green leaf foods, particularly kale, is perhaps much understated by the figure 7540 I.U. for raw kale in *Agriculture Handbook No.8.* Evidently this does not take full account of the larg numbers of determinations of vitamin A values of kale made by Dr. Todhunter in the chemical laboratory of Columbia University. Granted that it might be "undue particularism" to emphasise any one average from the work of Todhunter, the same upgradation is doubtless applicable in at least equal degree to some of the figures which did enter into the averages now appearing in the *Handbook.* For example: chard leaves, 8720; collards, 6870; dandelion greens, 13,650; parsley, 8230; spinach, 9420; turnip greens, 9540 I.U. per 100 grams of the edible portion.

Vitamin A Values of Typical Foods: International Units Per 100 Grams

Food	*Range Within Which The Final Average May Be Expected To Be found*
Green leaves such as kale, spinach, turnip greens	(10,000)a
Peas, fresh young green	600-1300
Mature seeds often	nearly negligible

Mill products and breadstuffs	negligible
Sugar and sweets	none or negligible
Fruits: Apples	40-100
Bananas	160-450
Cantaloupes	400-3500
Oranges	50-400
Vegetables: Broccoli	3000-9000
Potatoes	20-50
Sweetpotatoes	1500-7700
Tomatoes	500-1200
Beef muscle	10-50
Milk	160-225
Butter	3300-5000
Eggs	1000-2000

The distinctions among these figures may very probably be accidental, for kale which appears among the lowest of the completely green leaves in the *Handbook*, is absolutely among the highest in fact. The writer has therefore advised the tentative average figure of 10,000 which is very near the *Handbook* figures for spinach and turnip greens (probably the most used greens of the North and South respectively), and as affording a part of the upward revision which is clearly required in the case of kale.

The evidence now available on the green leaves utilised as food for man indicates that representative samples contain around 10,000 International Units* per 100 grams; but it should be insisted that similar high values are to be hoped only in leaves that are *completely* green. Leaves which are comparatively colorless must be expected to be of low vitamin A value. Thus Kramer and coworkers found less than one thirtieth as much vitamin A value in the white inner leaves of head lettuce as in the green outer leaves of the same plants. And the white inner leaves of tight-headed cabbages are of very much less vitamin A value than the green leaves of the loose leaf cabbages, while the still greener leaves of the closely concerned kale have much higher vitamin A values.

In 1927 Quinn, Burtis, and Milner exhibited the correlation with greenness in plant tissues other than leaves when they found snap or string beans and green peppers to be of high vitamin A value. Crist and Dye extended this to the case of green *versus* bleached asparagus tips.

The vitamin A values of other parts of plants are in common much lower; but in some cases vary extensively with the extent to which carotene is transported to, and stored in, few organs other than the leaf. Thus most roots and tubers are of very low vitamin A value, while the carrot with its high carotene content shows continuously a high vitamin A value, and in sweetpotatoes the value is high but varies extensively with the depth of yellow color. But not all yellow plant pigments are precursors of vitamin A; for instance, xanthophyll is not. However, F.L. MacLeod has displayed that in sweetpotatoes, and M.C. Smith has shown that in squash, the differences in depth of yellow color of the flesh are indicative of corresponding differences in carotene content and hence in vitamin A value.

We have already discussed, the leaves of forage plants and pasture grasses are of high vitamin A value. The extent to which this value is conserved in hay depends largely upon the curing process. Grains (and therefore our cereals and breadstuffs) are of very low vitamin A value.

As milch cows may consume grains and grasses in rather dissimilar proportions, a question arises as to the extent of the resulting variations in the vitamin A values of milk. In the long run the nursing mother or the lactating animal is relied upon her food for all the vitamin A value which she puts into her milk; but the storage capacity of the body for vitamin A permits a regulatory process by which the vitamin A value of the milk can, be maintained during enough periods of low intake by drawing upon the body stores accumulated in periods in which the intake was high. Thus the vitamin A value of the milk eventually relies upon, but does not literally vary with and as, the vitamin A value of the food.

The vitamin A values of sea foods develop much as do those of the land. Diatoms, small algae, and other small aquatic plants which have hence formed provitamin A may then be eaten by small crustacea, these by fish, and these in turn by larger fish such as the cod. Thus the cod, through propably two or three or more intermediaries, has received its vitamin A from sources finally analogous to those which supply the cow.

The U.S.P. standard for cod liver oil calls for not less than 850 units *per gram,* and halibut liver oil is much richer in vitamin A, having, in comformity wise to McCollum, about 20,000 units per gram, while percomorph oil is still richer.

Storage of Vitamin A

The combination of properties which, formulates vitamin A capable of storage in the body, in quantities enough to meet its nutritional requirements for comparatively long times, is of great importance to nutational well-being and health at all ages, and the clear recognition of the importance of this factor in the nutritional economy is highly necessary to an sufficient appreciation of the true magnitude of the intake of vitamin A values.

At some times and in some quarters, there has been failure of due recognition of the significance of the vitamin A value of the food supply, due to of the very reality that bodily storage may protect from prompt penalty in periods of subsequent shortage.

Both for interpretation of experiments and for the everyday application of our new knowledge, we must keep in mind that the body does not rapidly eceive nor rapidly lose the full store of vitamin A which it is capable of taking on when habitually provided a liberal intake. Cammack (Sherman and Cammack, 1926) found this many years ago in experiments with moderately liberal daily intakes, and the same reality appears to have been accountable for some confusing observation after Wald's administration of colossal doses. Conceivably it may also have a bearing upon lack of pronounced correlation between bodily stores of vitamin A and either blood values or light-adaptation data in the work of Steininger, Roberts, and Brenner (1939).

Even at weaning time young animals (doubtless comprising babies) may already have a large store of vitamin A in the body. The body can also store vitamin A at subsequent ages.

In one investigation (Sherman and Boy ton, 1925) laboratory animals of known nutritional history were killed and dissected and graded weighed amounts of their tissues were fed as the only ways of vitamin A to young experimental animals. In this source it was shown directly that there is an actual storage of vitamin A in specific tissues, and it became possible to study quantitatively the distribution of the stored vitamin in the body and also to condition directly the extent to which the vitamin A content of the body is affected by the intake of this vitamin in the food. If adipose tissue and skin be ignored, at least nine tenths of the total vitamin A in the body of a well nourished adult rat is found in the liver weight for weight, kidney appeared about 40 times, lung above 40 times, and liver between 200 and 400 times as rich in this vitamin as the muscle of the same animal. Moderate differences in the vitamin A content of the food, like are well within the range of variation perhaps to be encountered in human experience, resulted very larg differences in concentration of this vitamin in the liver, and distinct differences in the amount of it found in lung tissue. That the vitamin content of the lung tissue is dependent upon the abundance of this vitamin in the food is of interest since it has also been found that the higher intake of the vitamin is accompanied by a lower incidence of lung disease. Baumann, Riising, and Steenbock (1934) have also studied the body storage as affected by nutritional intake and the distribution in the body. They assigned about 95 per cent of the body's vitamin A content to the liver.

In another study of accumutation of vitamin A in the body (Sherman and Cammack, 1926) the placing of animals upon diet devoid of vitamin A was preceded by the feeding for different lengths of time of a diet bountifully supplied with the vitamin. This resulted in very great prolongation of life upon the vitamin-A-free food, showing that large surpluses of vitamin A can be stored under these conditions also. The maximum storage of vitamin A in rats of dissimilar ages was closely approached by feeding a ration containing 4 per cent of cold liver oil, which supplied about 80 times as much vitamin A necessary for sufficient nutrition. However, the store does not enhance in arithmetical proportion to the richness of the diet in vitamin A, since the animal uses the vitamin given in the food more freely or less economically as the maximum store is approached.

The storage of vitamin A in the body has been to be an important factor, it appears correct that the supposed lesser requirement for this vitamin by the adult than by the young is apparent readily than real, and largely attributable to the reality that the adult has had more opportunity to give himself with a bodily store of this vitamin which then carries him over periods of deficient intake.

Storage of Vitamin A in the Body May Complicate Interprtation in Research

Vitamin A has a combination of physical and chemical properties which formulates it capable of long storage in the body, and up to qualities that are large regarding to the body's daily need. It is to our nutritional benefit that we can carry in our bodies similar a valuable asset but its magnitude may be under-valued unless it is kept in mind that the storage factor formulates long term experimentation especially important regarding to vitamin A.

A distinguished pediatrician in the early history of vitamin A endeavoured the experiment of feeding one group of children in an institution skimmed instead of all milk for 9 months, which he judged would make it a readily thorough test, covering the comlete winter season. Finding that this group ostensibly went through the winter as well as the others, he drew a final opinion which doubtless underestimated the importance of vitamin A for children; as he did not percept the importance of the storage factor, and that a nine months' experiment with children was only the equivalent of about a nine day experiment with rats.

Innumerable experiments exhibit that the body can store vitamin A at all ages. The very great extent to which bodily storage is influenced by the level of intake in the case of this vitamin is also now well established.

The Position of Conversion of Carotene to Vitamin A

In 1931 following the work of Moore and of Olcott and McCann the site of conversion of carotene to vitamin A was usually accepted to be liver. Conflicting consequences were reported by other workers but inconsistencies were attributed, at least in part, to species differences.

In the investigation of this difficult there has been renewed activity. In 1946, Sexton, Mehl, and Deuel advised that the intestine might be the site of the conversion, later work in their laboratory showing that in the rat this is actually the case.

At the same time other research groups, comprising Glover, Goodwin, and Morton; and Thompson and Kon in England, using different techniques and, in addition to the rat, various herbivora and pigs and chicks, also arrived the final opinion that the intestinal wall is the site of conversion of carotene to vitamin A, the efficiency with which the modify occur varying in dissimilar species.

Vitamin A : Its Absorption and Transport

It was observed by Bill, Drummondond palmeer In 1935 that the vitamin A content of the lymph increased after vitamin A or carotene had been ingested, the vitamin A in the lymph being esterified Popper and Volk (1944) found vitamin A in the facteals of the rat after oral during with vitamin A, absorption occuring from the small intestine, most quiucly between the upper and middle thirds. After oral dosing with carotene, Goodwin et. al. found the vitamin A level in the lymph of herbivora increased, and that no carotene appeared in the blood. Apply a fluorescent method, Radice and Herraiz (1947) found in the rat that obsorption of vitamin A took place into the lymphatics, most rapidly from the jejunum. Eden and Sellers found marked increase in the vitamin A content of intestinal lymph, especially from the duodenal region, after offering bullocks, sheep, and rats doses of the vitamin, but there was no greater modify in the content of either portal or systemic blood, or of lymph from other regions of the body. Further evidence for the lymphatic route of transport of vitamin A was given by the work of Goodwin, et al. (1948), in which increased levels of vitamin A were found in the lymph accumulated by cannula from the intestinal region in goats after feeding either vitamin A or carotene while blood values remained unchanged.

The work of Eden and Sellers confirm the previous observation by Gray, et al. in 1940 that verbally ingested vitamin A ester is hydrolyzed in the intestinal tract, the alcohol clearly being the form in which it is absorbed; but it is re-esterified in its passage through the intestinal wall and seems in the lymph all but completely as the ester. It is also as the ester that it seems in the circulating blood.

Various Stages of Metabolism of Vitamin A and Carotene

Conversion of Carotene of Vitamin A. After it was that carotene is a way of vitamin A the next step was to find out where in the body the conversion of the pigment to the vitamin occurs. It was belived that the conversion occur in the liver. It was suspected due to of the high vitamin A content of the organ. The hypothesis was questioned, however, when the provitamin administered intravenously was not trandsformed to the vitamin. Further doubt was cast on that belief when blood vitamin A was elevated after feeding carotene to rats, with the liver taken out of circulation. The liver was cut off from all connections by trying off the veins, arteries, and general bile duct. It was under the conditions of this experiment, along with other evidence, that the intestinal wall was found to be the chief site of the transformation in the rat.

The liver and other tissues have been displayed to be alternative sites for the transformation of some carotene to vitamin A. Since man is capable of absorbing carotene from the intestine, other sites for transformation may be of ineater significlance. Education is needed about the site and the conversion mechanism of the carotenoids to vitamin A.

The amount of conversion of carotene to vitamin A is variable and depends on the dietary way and efficiency of absorption. Pure crystalline beta-carotene was found to be more efficiency used than beta carotene from the green vegetables in studies with adults. In average diets carotene is estimated to be one-half as well used as performed vitamin A: 3000 IU of carotene in the diet would be expected to produce only 1500 IU of vitamin A in the body.

Absorption. Vitamin A and carotene are absorbed from the small intestine into the lymph system. The maximum absorption is arrived three to five hours after consumption. The rate of absorption of vitamin A is more quick than that of carotene.

In the small intestine the absorption of both vitamin A and carotene is influnced by other substances. Bile, dietary fat, antioxidants, and detergents enhance absorption. Bile is. necessary for the absorption of carotene. Mineral oil may interfere with absorption, especially if taken to gether with food, or if a large amount of oil is fed with a less than sufficient intake of the vitamin. Both vitamin A and carotene are soluable in mineral oil; since the oil is not absorbed, some of the vitmain A and provitamin A are eliminated with it. Under specific conditions, however, the oil is found not to have a deleterious effect. In a study with adults one ounce of mineral oil taken at bed time was found not to weaoen the absorption of either vitamin A or carotene.

Storage and Excretion. In the human being about 95 per cent of the vitamin A stored in the body is found in the liver with little amounts in the lungs, body fat, and kidneys. The amount is least at birth and increases with age. The store of vitamin A in the liver depends on the quantity in the diet and the amount absorbed. It has been estimated that a normal liver may contain as much as 600,000 IU of vitamin A, sufficient to supply vitamin requires for one year.

Vitamin A is excreted by source of the intestine. It has been absorbed that the fat-soulble vitamins are excreted by the fecal pathway, the water-soluble vitamins by the urinary pathway.

Valuation of Vitamin A status

One of the early symptoms of vitamin A deficiency is night blindness or subnormal dark adaptation. Many instruments similar in principle are used for the quantitative measurement. Among them is the photometer (adaptometer) (Fig.5), which measures the amount of light (intensity) needed for the subject to perceive objects privious and after his exposure to bright light. The intensity of light and the time needed to regain vision after bleaching of the visual purple are indicative of vitamin A status. Signs of deficiency are recognized by the requirement for a greater intensity of light and a longer period of time for the regeneration of the visual purple.

Different ways of error are present in the photometer procedure, both in the instrument and in the subjective response of the person being tested. Conditions other than that of vitamin A deficiency may impair dark adaptation. If quice and distinct improvement follows the ingestion of vitamin A, it is proved that a lack of the vitamin was a causative factor.

The blood level of vitamin A and carotene bear a relationship to intake and are indicative of vitamin A status within specific limits. Adults placed on a vitamin-A-free diet are found to display no modify in the level for several weeks, followed by an ultimate lowering in the concentration. A carotene-free diet consequences in prompt reduction of the blood carotene content. The body has stores of vitamin A that can be mobilized for use, whereas carotene is not stored.

Deficiency of Vitamin A in the population of this country seems to be comparatively rare. The strength of the body to store vitamin A provides for an effective emergency supply.

Hypervitaminosis A

Excessive consumption of vitamin A may lead to taxic symptoms in both man and animal. Hypervitaminosis A is being reported increasingly in both children and adults who have taken excessive dosages of concentrates of the vitamin for diverse lenghts of time. A case report of a 39-year-old woman who was taking 150,000 IU of vitamin A daily in expectation of improving her vision has been illustrated. The symptoms were swelling of the feet and ankles, fatigue, lassitude, weight loss, and intermittent pains in the shoulders, wrists, and knees. Discrete superficial hemorrhages in the retinae of both eyes in a 17-year-old boy who had been taking 200,000 IU of vitamin A daily for 18 months for acne was reported. Arctic travellers who have consumed polar bear liver, which is very high in vitamin A content, have manifested symptoms of acute hypervitaminosis A.

The signs of acute toxicity are not simlar in all cases; most usually observed are skin lesions, an itching rash in some instances, in others a coarsening of the skin, thinning of the hair, and repeatedly pain in the joints, with hemorrhage between the bone and its covering. These same symptoms may arise from other causes. A back ground of consumption of excessive amounts of the vitamin and improvement in physical condition with the discontinuance of its utilize led to ultimate authority that hypervitaminois A was the liable for the trouble in recognized cases. Acute toxicity may occur is an end stage of prolonged chronic toxicity resulting from overdosage over a period of time. The toxic dose is highly variable among individual particularly infants and young children, but is considered to be over 100,000 IU daily. III impression from lesser amounts administered over a period of time have been reported. The Council on Foods and Nutrition of the American Medical Association state that "there is definite possibilities of harm from the prolonged ingestion of vitamin A in excess of 50,000 USP units daily." The utilize of vitamin concentrates that reqjuire only a few drops for a dose afford an easy means of giving above is desirable. For instance, around 50 drops of many of the general concentrates were found to produce one-half teaspoon, providing in that small measure 15,000 IU of vitamin A. Concentrates of the vitamin are incorporated in multiple vitamin capsules in varying amounts, some contain 5000 IU per capsule, some 15,000 IU and others 25,000 IU and 30,000 IU.

Vitamin A : Its Health Significance

The relations to health are clearer than the relations to disease. Comparativety early in the history of our knowledge of vitamin A, Dr. F. L. MacLeod found that in sufficiently long-term experiments ample differences developed in family life histories of rats fed wheat and milk mixtures according as all milk or skimmed milk was used (Sherman and MacLeod, 1925). The smaller mount of vitamin A (supplemented, no doubt, by reserves in the bodies of the animals at the nception of the experiment) was enough for normal development up to about average adult size. And the animals presented an excellent appearance during their youth and into adulthood. But this diet did not support successful reproduction, nor did the low-vitamin-A animals have normal longevity, as did those obtaining more of this vitamin. In these early experiments, only two levels of vitamin A feeding were used, and these so far apart that the eventual difference of outcome was too drastic for strictly quantitative measurement.

Dr. E. L. Batchelder (1934) fed animals in similar source in other regards upon four levels of dietary vitamin A and found that this stepwise increasing of vitamin A intake in the proportions 1, 2, 4, and 8 induced similar improvements in nutritional performance and family histories. The gains from extra feedings of vitamin A are summarized in Table 2. The number of animals were, however, not amply large to allow of any quantitative comparisons of results. In subsequent experiments at Columbia alive comparisons were made in larger numbers by Dr. H. L. Campbell and Mrs. H. Y. Trupp.

Here the basal ration was the Columbia Diet A (Laboratory No. 16) on which rat families have thrived throughout 70 to 75 generations so there is no doubt that the diet was enough to all the nutritional requirements of the experimental animals. This basal diet had 31.U.* of vitamin A per gram of the dry food mixture, certainly sufficien as just explained. When this allowance of vitamin A was doubled the consequence was a superior record comprising a longer average life. And when the vitamin A allowance was again doubled, formulating a fourfold increase over the original (about minimal-adequate) intake, the health record and length of life were still further bettered. Thus vitamin A is one of the outstanding nutrients of which stepwise increased intakes induce correspondingly higher health and prolong life up to intake levels about 4 times that of minimal adequacy. There is no doubt that in these experiments the addition of extra vitamin A to a diet which already have "enough of everything" did positively and constructively build higher health and longer life. And we require have no doubt that this same thing may usually occur with people.

The enhancement in the average length of normal adult life in these experiments was 10.4 per cent with males and 12.1 per cent with females. If one finds doubt about side questions of concerning of vitamin A to disease, one can still be confident that it certainly promotes positive health.

Table : *Effects of dissimilar Levels of Intake of Vitamin A (Based on work of Batchelder)*

In Take	*Growth And Size*	*Adult Vitality*	*Length of Life*	*Vitality of Young*
At level of :	Within normal range	Within normal range	Within normal range	Within normal range
Increased from 1 to 2	Slightly increased (within normal	Increased (within normal range)	Increased (within normal range)	Distinctly Increased range)
Increased from 2to4	Slightly increased (within normal range)	Slightly increased	Slightly increased	Distinctly increased
Increased from 4 to 8	No further change	No directly measurable change	Slightly increased	Distinctly increased

How human beings may best benifit by what has been learned through animal experimentation with different intake levels of vitamin A value, is a difficult worthy of careful thought. And, to guide our thinking, we now have much more clearly and firmly established scientific research than existed even a few years ago, or than most people are perhaps yet completely aware of.

This animal experimentation upon the responses of the health and life history has been chiefly with the rat due to of its close nutritional likewise to the human. There is no cause to doubt and good cause to suppose that, in human as in rat nutrition, best consequence in long-term experience will requare at least 2 to 4 times as much vitamin A value as suffices for minimal adequacy. If the minimal-adequate amount is taken as about 3000 I.U. for normal adult maintenance, from 6000 to 12,000 would be a scientifically more logical allowance (than the 5000 I.U. allowance of the National Research Council) to give both for individual variations in requirement and for the maintenance of similar a bodily reserve as has been found to be favourable to higher health and longer life.

To the extent that the vitamin A values of dietaries depend largery upon the precursor carotenes readily than on vitamin A itself and the doubts as to the completeness of avilability of these precursors in digestion and assimilation, one will opt to have his intake of vitamin A value nearer the top than the bottom of the 6000-12,000 I.U. zone more than suggested.

Also it is scientifically sound in principle to give liberal intakes of a nutrient which seems to be included in a number of dissimilar functions and in so several major and minor ills as is vitamin A, even if some of these involvements are not completely clear.

The common principle, that nutritional shortages not of enough severity to be unspicuously diagnosed may nonetheless be reasons for dietary improvement, is ostensibly worthy of particularly careful observance in the case of vitamin A values. This is a scientifically studied discrimination; not only an undiscriminating open-handedness with food.

Suggested Permission for Vitamin A

The Food and Nutrition Board Recommend Daily Allowances of vitamin A at levels said to be "adequate to meet the requirements of health. They are calculated to be about twice the minimal need. The recommendation for young men and women 18 to 35 years of age is 5000 IU, the same daily allowance is recommended for older adults. The allowance is made on the ground that 4000 IU are yielded in the average diet by carotene and 1000 IU by vitamin A. Since carotene is only about one-half as well used by the body as is performed vitamin A, if all the vitamin A activity in the diet were from carotene the allowance would be 6000 IU instead of 5000 IU. On the other hand, if all the vitamin were given by the performed vitamin the allowance would be 3000 IU instead of 5000 IU.

Vitamin A : Its Sources

Performed vitamin A occurs only in foods of animal origin; carotene is available in both plant and animal products. The most important ways of vitamin A are fish-liver oils, liver, butterfat, fortified margarine, and egg yolk. The concentration of vitamin A in the liver increases with age and varies with the species. The potency of butterfat and egg yolk are influenced by the diet of the cow and the hen, respectively. In the spring and summer when green grass becomes readily available, the vitamin A content of the milk and eggs produced by the animals is higher. However, the improved and enriched livestock feed used today formulates possible the provision of an ample way of vitamin A in animal rations the year round.

The most petent ways of the provitamin are the yellow, yellowishred, and green vegetables and fruits. The yellow carotene in the green products is masked by the presence of chlorophyll. Among the best ways of carotene are carrots, sweet potatoes, apricots, spinach, tomatoes, and kale. Some vegetables and fruits deviate from the common color rule; even though yellow in color, oranges, rutabagas, and yellow wax beans are not appropriate sources of carotene. Head lettuce and avocados, two green-colored foods, give small amounts of carotene. Foods comprising little vitamin A are nuts, grains, vegetable oils, muscle meats, and light-colored fruits and vegetables. In most of the foods one-half or less of the carotene is converted to vitamin A. As a result, the carotene content of a food is by two in most examples to obtain the biological activity in in connection with vitamin A.

The vitamin A values of some general foods are provided in Table 1. The term vitamin A *value* of foods is utilised instead of vitamin A *content*. It would be not probable to point out that carrots, for instance, contain 12,000 units of vitamin A per 100 gm; actually they contain no vitamin A at all; they are only rich in crotene content.

In compiling the tables of food composition published as Agriculture Handbook 8, Wait and Merrill tramsformed ug of carotene in foods to IU of vitamin A activity on the basis that 0.6 ug of beta-cartone, and 1.2 ug of the sum of the other carotenes with vitamin A activity or of cryptoxanthin are equivalent to 1 IU of vitamin A value.

Stability of Vitamin A and Carotene in Foods

Since both vitamin A and carotene are not soluble in water and stable to ordinary cooking temperatures, there is little loss in food preperation. Canning and freezing reason little loss, but drying and dehydration consequence in mofor destruction through oxidation.

The growth of rancidity in fats leads destruction of both vitamin A and carotene. Storage of fats in a cool, dry place delays the onset of rancidity and hence supports to preserve the vitamin. Magarines were found to lose than 25 per cent of the original vitamin A content when stored at 14° F for a period as long as 2 years, but when stored at 64° F, 75 per cent of the original vitamin A content was lost in only 17 weeks

Meeting the Daily Require for Vitamin A

Dietary records of college students often display an intake of vitamin A less than the Recommended Allowance. An evaluation of these dietaries shows a low intake of dark green and yellow vegetables. It is estimated that vegetables and fruits give about 60 per cent of the available supply of the vitamin A value in the United States. The incorporation daily or three to four times a week of a serving of a dark-green or yellow vegetable is an easy source to lay the foundation for enough intake of vitamin A. Table 2 displays groups of foods for meeting the daily vitamin A need.

Individuals vary in their ability to absorb carotene and in its conversion to vitamin A. Therefore, there should be enough intake of carotene to permit for any discrepancy. Foods providing vitamin A, like eggs, liver, and cheese, should be incorporate.

The Effect of a Deficiency of Vitamin A

Xerophthalmia

It is a condition of the eye resulting from severe deprivation of vitamin A. The tear glands do not to function, the eye is dry, the cells of the *cornea* (the candid outer covering of the iris and pupil) become dry and in thought full cases ulcreated. If the condition is not arrested by provision of vitamin A, blindness ensues. The cornea, the part of the eye suffering the severe damage in xerophthalmia, is an epithelial tissue. The condition occurs in both man and animal. Xerophthalmia rarely occurs in the United States. It is, however, general occurrence among the young in China. It is a reas on of blindness among the rice and cassavaeating people of the southern part of India and Ceylon and is found in Indonesia, Malaya, the Philippines, and in parts of Central and South America. The most of the cases occurs in children two to six years of age. In 1914-1918 children in Denmark were afflicted; butterfat was shipped from the country, and skimmed milk comprising little vitamin A used. The condition can be developed in experimental animals deprived of vitamin A.

Composition and Functioning of Visual Pigments

The retina of the eye comprises visual pigments that are a combination of vitamin A and a protein. The retina is the light-sensitive layer at the back of the eye; it may be likened to the sensitive film of a camera on which the image occur. Two types of receptor cells are in the retina—the rods that function in vision in dim light, and the cones that function in bright light and color vision. The terms "rods" and "cones" arise from the shape of the cells. Rhodopsin is a vitamin A-containing pigment in rods, iodopsin in cones.

Growth

Growth of young rats to grow when they are deprived of particular foods was the first symptom that led to the discovery of vitamin A. In the early studies privious vitamin A had been isolated and chemical methods found for its determination, development rate was the grade of measurement of the vitamin A potency in a food. Animals maintained on a vitamin-A-free diet will cease growing when the body is depleted of its vitamin stores. Materials to be conditioned for vitamin content were then fed in known amounts for a prescribed period of time and the development response noted. Better development was associated with higher vitamin potency in the food. If vitamin A in enough amount is not given for normal growth, the bony structure will suffer a development failure privious the soft tissues manifest the stunting effect. The cessation of bone development may reason overcrowding of the brain and central nervous system. It is now clear that the paralysis sometimes accompanying vitamin A deficiency in very young animals is since of failure of bone growth, that reasons cranial pressure with consequent brain and nerve injury. In some escamples there is pinching of the optic nerve, resulting in blindness.

Health of the Epithelial Tissues of the Body

The outer surface of the body is covered with epithelial tissue, the internal cavities of the body lined with it. The tissues are differntiated to better serve their individual functions; the external covering is the resistant, protective epidermis; the internal tissue is a secretory mucous membrane. Vitamin A's deficiency A reasons an abnormal condition of the epithelial tissues. The specialized functions of the tissue are suppressed, and it is cenverted to a keratinized (dry, horny) kind of epithelium. Excessive dryness of the skin may consequence from an inadequate intake; the mucous membrane may fail to secrete normally and becomes less opposite to bacterial invasion.

Vitamin A and Vision

From the standpoint of the chemistry of nutrition, the outstanding relationships of vitamin A to vision are: that vitamin A is essential to the regeneration of visual purple, the bleaching of which in the retina is one of the chemical reactions included in vision; and (2) as a consequence it follows that the ability to see in a dim light, or to adapt to a changed intensity of light, is very directly related to the abundance and immediate availability of the supply of vitamin A. The mechanism of vision is also two-fold: that of the cones of the (diagrammatic) physical structure of the retina.

Having used visual adaptation tests Jeans and coworkers (1936), found seriously frequent evidences of shortage of vitamin A among school children, particularly in low-income families. subsequent on much attention has been provided to the improvement of instruments for utilize in similar tests, to the critical study of details of technique and interpretation, and to the extension of this common method both in surveys of populations as to their nutritional status and in laboratory research upon the quantitative difficulties of vitamin A requare in human nutrition.

Whether measurement of visual adaptation or of the concentration of vitamin A in the blood will prove the more valuable criterion is a research question, the answer to which will doubtless rely both upon the growth of techniques and upon fuller and more critical knowledge of the nutritional importance of vitamin A. It is perhaps, that a real shortage of vitamin A in the body will inevitably impair the efficiency of visual adaptation; but also that similar impairement may sometimes consequence from other causes. And since bodily storage may be similar a large and imperfectly comprehended factor, the recent dietary and therapeutic history may not forever permit a sound judgement as to whether or in what sense a real shortage exists.

Vitamin A and Metaplasia

The concerning of vitamin A to mucous membranes is a reality to very far-reaching importance. It has been found to be important in the eyes, in particular secreting glands, in the respiratory system, the alimentary canal, and the urinary tract; and it is easily conceivable that still other similar relationships may remain yet to be found out.

Mori, Bloch, and Blegvad independently mentioned diseased conditions of children which in the light of all authority now available seem to have been due primarily to shortage of vitamin A. While xerophthalmia was the symptom which called attention to these cases, it was noted that there was also a high incidence of respiratory disease which was most likely because of the weakening of the respiratory system by the same dietary deficiency. Blegvad argues that the deaths which occured among the children who displayed the eye trouble were due in most of the cases to respiratory disease.

Experimental Deficiency in Laboratory Animals

In experiments with laboratory animals the readily observable sign of vitamin A deficiency is the growth of an ophtalmia (xerophthalmia, keratomalacia, conjunctivitis). This was first recorded by Osborne and Mendle in 1913 and has occurred so always (at least in the experience of most observers) as to be considered as characteristic result of a lack of vitamin A. But it is only one of several results, and not the most potent. Further studies have exhibited that vitamin A deficiency consequences also in an increased incidence of respiratory disease; of skin, car, and sinus infections; of inflammations and infections of the alimentary tract; and even of renal calculi (Osborne and Mendle, 1917). The underlying reason of the weakness that may show itself in so several sources is the histological modify that Wolbach, Howe, and Church have mentind as follows: "The certain effect of the scarcity of fat-soluble vitamin A in albino rats, guineapigs, and humans is found in epithelial tissues. This influence is the substitution of strarified keratinizing epithelium for the normal epithelium in many parts of the respiratory tract, alimentary tract, eyes and paraocular glands, and the gennito-urinary tract."

The greater incidence of infection observed among individuals on diet of low vitamin A value may be since of the displacement of normal by squamous epithelial tissue implies not only a loss of local secretion but also a loss of the *cilia* that act with the secretion in cleansing the surfaces of the respiratory tract by expulsion of foregin particles. Plainly the normal surface hence tends to protect the respiratory system both from bacteria and from miscellaneous particles that might reason mechnical injury and so enhance the danger of the setting up of infection by similar bacteria as may remain. It has also been suggested that the local roughness and stickiness which accompany the breaking-up of the normal tissue (when it is first being replaced by the squamous) may enhance increase the danger of lodgment of bacteria which would usually have been expelled.

Thus even our present incomplete knowledge shows above one source in which shortage of vitamin A may increase the dangers of infectious disease. Whether it particularly increases susceptibility, in the technical sense in which bacteriologists and patholohists are apt to utilize this term, we do not know. Emphasis tends to be laid upon the characteristic metaplasia of epithelial tissue which several investigators have observed to consequence from deficiency of vitamin A and which structurally weavens what the *journal of the American Medical Association* has called "the bodys first line of defense." This metaplasia of the epithelium has been observed in many parts of the body and in every species which has been thoroughly studied. It most likely underlies the external eye symptoms as well as the other ills more than mentioned. In Tildern and Miller's experiments upon monkeys, a scarcity of vitamin A resulted in injuries to the digestive tract which were usually fatal privious the growth of a conspicuous degree of ophthalmia.

The ascendancy of vitamin A, unquestionably potent at the first line of defense" against infections extends also beyond this, has been displayed by many investigations (among others those of Boynton and Bradford, 1931; of Lassen, in 1932; and of McClung and Winters, 1932), in which infective agents have been *injected* under experimental control, and it has been found that the animals more liberally supplied with vitamin A survived better.

To getrid of confusion, we should keep in mind that there are wide differences in access among those presumably competent to discuss the concerning of vitamin A to infection;

and that these differences are to a considerable extent evaluated by the reality that anthorities vary in their sources of access to the difficult those who have sought to connect vitamin A with the immunological mechanism as commonly conceived have reported negatively. Those whose access is somewhat extensive but who still concentrate upon the inquiry for clearly direct primary relations, tend to report either negatively, or inconclusively. But on the more extensive (if technically less precise) question, there is much positive evidence that the level of intake of vitamin A does (whether directly or indirectly, primarily or secondarily) ascendancy the frequency, or severity, or duration, of infectious disease—not particularly of any one infection alone, and also not equally of all infections.

Both age and prior feeding may ascendancy the consequences of a vitamin-A-poor diet. In a typical succession of observations, three fourths of the rats placed upon vitamin-A-free food at the age of about one month (end of infancy) developed the characteristic ophthalmia privious death; but developed in only about one fourth of the rats that were from two to nine months old when subjected to the same dietary deficiency (Sherman and Storms, 1925).

The older rats, however, proved more vulnerable than the younger ones to lung infection, the frequency of which, in Columbia experience, such as that of Steenbock and other investigators, is enhanced by diet deficient in vitamin A.

The difficult of Human Requirements and Adequancy of Food Supplies

The National Research Council's Recommended Daily Allowances for vitamin A values of dietaries, expressed in International Units (I.U.) are as follows—

Requirments of Vitamin A

Men (70 kg.),	5000 I.U.;
Women (5b kg.),	5000 I.U.;
pregnancy, latter half,	6000 I.U.;
Lactation,	8000 I.U.;

Children under—

1 year,	1500 I.U.;
1-3 years,	2000 I.U.;
4-6 years,	2500 I.U.;
7-9 years,	3500 I.U.;
10-12 years,	4500 I.U.;

Girls—

13-15 years,	5000 I.U.;
16-20 years,	5000 I.U.;

Boys—

13-15 years,	5000 I.U.,
16-20 years,	6000 I.U.

These allowances are materially higher than earlier estimates of the amounts required for the prevention of similar signs of deficiency as have been mentioned above, and at the time of adoption early in 1941 were considered as carrying liberal margins more than actual requirements. Soon afterward, however, Kruse (1941) by utilize of a more delicate method, found authority of a much greater prevalence of lack of vitamin A than had previously been known or suspected.

Bitot had mentioned the occurrence of elevated spots on the conjunctival membrane of the eye, and these had become known as Bitot spots, but were not usually as significant until Kruse observed repeatedly that they disappeared-although sometimes very slowly-under vitamin A therapy. He therefore considered Bitot spots a indication of vitamin A deficiency (*avitaminosis* A), and proceeded to a biomicrosopic study of the parts of the eye where they are most often seen. This investigation exhibited that the typical Bitot spot, visible to "gross" examination, is a relatively advanced sign, the previous or lesser stages of which can be seen only with the aid of biomicroscopy but are then found to be of more frequent occurrence than the grossly visible spots.

If the evidence of this more delicate method be taken as conclusive, it follows that a large proportion of the people previousrely considered as amply nourished had actually received too little vitamin A, and that the true human requirment for vitamin A is considerably higher than hitherto supposed. But not all nutritionally minded physicians yet regards these minute and yet unknown or disregarded signs as ultimate evidence of vitamin A deficiency.

At present, therefore, in conformity with one's judgement of these biomicroscopic observations, one may 'respect the Recommended Allowances of vitamin A value as liberal or as scanty from the point of view of preventing deficiency signs.

Similarly one may realize that the average American dietary does or does not carry a completely satisfactory margin of vitamin A value. There appears to be no doubt that the actual vitamin A value of the American food supply has been completely maintained, if not increased, during recent years, due to better conservation of food (including fishery) resources and enhanced production of green and yellow vegetables and tomatoes in home gardens. These improvements readily can, and wisely may, be carried further, for medical experience and animal experimentation have clearly displayed that very liberal margins of vitamin A value in the dietary provide added benifit in the long run.

Mellanby has reported, from his combined hospital experience and animal experimentation, that increased allowances of vitamin A (or precursor) continue to eliminate the dangers of infection up to levels of intake four times that needed for normal nutrition in absence of exposure.

For people living under everyday conditions and not encountering particular exposure to infection, perhaps the best evidence as to long-run desirability of a liberal intake of vitamin A value is that of the more than cited full-life and successive-generation experiments with laboratory animals. These experiments displayed that successively increased levels of vitamin A intake resulted in successively improved life histories and vigor of offspring, up to diets of about four times the vitamin A value which ordinary criteria would accept as sufficient.

Importance of Vitamin A

Vitamin A being used by the body to maintain the health of the skin, teeth and mucous linings of the respiratory organs is required particularly in illness in which these are involved. To guard against them, supplies in the daily diet should be kept up to the standards given. In treatment higher doses are competent in pharmaceutical preparations can be prescribed. Dazes up to 100,000 or 200,000 i.u. are sometimes given daily.

5

Vitamin B: Its Nutrition

The requirement for vitamin B_1 varies in conformity with the amount of food eaten and the energy expended in movements in all sorts. As muscular activity, which relies on the oxidation of carbohydrates, continues throughout life, vitamin B_1 must be always supplied. Besides, this vitamin differs from vitamin A in not being stored for long periods. In general, the fat-soluble vitamins tend to be stored in the body, but the water-soluble vitamins, i.e. those of the B group and vitamin C, are not stored. If more is eaten than required it is radically excreted again in the urine. Since vitamin B_1 is used day by day, as it is consumed, it has been much easier to find out property what the daily needs are in fact, as early as 1934, Cowgill was skilful to suggest a mathematical formula to express the daily requires regarding body weight and calorific intake.

The B Vitamins

The first step was the recognition that vitamin B was not

a simple materials but comprised of at least two factors. These were differentiated first as one fraction unstable to heat and another stable. The heat-labile fraction became vitamin B_1 and was accepted as the anti-beriberi or antineuritic factor; the heat stable fraction was B_2, known in America as vitamin G.

Vitamin B_1

Jansen and Donath (in Denmark) in 1926 extracted a highly potent vitamin B_1 fraction from rice polishings and were able to obtain this in crystalline form.

Their work led to other try at concentration and it was found that from an aqueous extract the vitamin could be concentrated and dried on divided powders like acid clay. This process was utilised in the manufacture of the first vitamin B_1 concentrates and the resulting powder (Pulv. vitamin B_1) was provided a place in the British Pharmacopoeia. It was similar a powder as this which was utilised as the first international standard for vitamin B_1.

From this stage to the synthesis of vitamin B_1 comprised much chemical research, first not the discovery of the formula of the vitamin and then into methods of building this up in the laboratory and it was nearly ten years privious vitamin B_1, known chemically as aneurin, or thiamin, was synthesized in 1936.

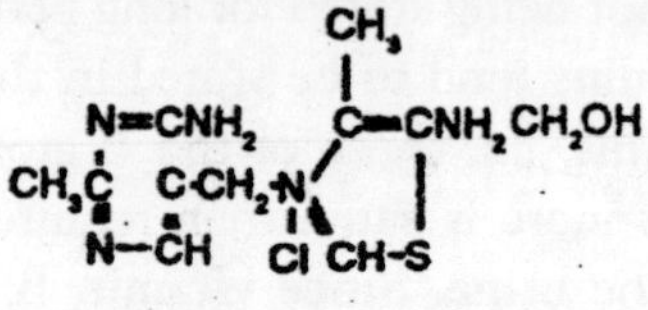

Formula of Vitamin B_1

It include nitrogen and sulphur in addition to carbon, hydrogen and oxygen, and its commonest salt is the hydrochloride.

The discovery of thiamine resulted from work along two sources of research: laboratory experimentation in normal nutrition, and medical studies of the disease beriberi.

Prof. Frederick Gowland Hopkins of Cambridge University in 1906reported feeding experiments with rats which showed the requirement in normal nutrition of some unidentified organic materials or substances soluble in water and in alcohol. Seeking identification of this factor, Hopkins postponed publication of the details of his experiments until

1912. Meanwhile, Dr. Casimir Funk, a Polish chemist working in the Lister Institute (London), had postulated the existence of a factor, the dearth of which reasons beriberi, and coined for it the name *beriberi vitamin* (e). Also he predicted (as had Hopkins) that other diseases might be found to be following lack of analogous substances.

Thiamine

Beriberi, a nervous system disease that is caused following lack of thiamine in the diet, has occurred in the Far East for centuries and has resulted in a high morbidity and mortality rate among the peoples of these countries. Not until 1882 was it displayed by Takaki, at that time Director General of the Medical Department of the Japanese Navy, that beriberi could be prevented in man by the intake of a good diet. last 10 years afterwards the observations of Christian Eijkman, a Dutch medical scientist in Java, began the research that ultimately led to the identification of thiamine as the anti-beriberi vitamin. Eijkman observed that the chickens housed in the hospital yard near his laboratory developed a disorder, that was resemble to human beriberi, when fed a ration only of polished rice. Following of the common paralysis and head retractions which occurred in the birds, he called this disease *polyneuritis.* He fund that the symptoms of the polyneuritis disappeared when rice polishings, the outer layers and embryo of the rice kernel, were fed to the birds. In 1911 the Polish biochemist, Funk, prepared a small amount of a crystalline material from a rice bran extract which he found cured experimental beriberi in birds. Since he called this materials *beriberi vitamin,* he deserves credit coining the word vitamin. Four years subsequent McCollum confirmed the existence of a single materials which not only cured polyneuritis but also encouraged development and he named this factor *water-soluble* B. Jansen and Donath, at Eijkman's laboratory in Java, demonstrated the vitamin's curative influence in human beriberi in 1926, after they had prepared enough crystals of the vitamin from rice bran. After ten years R.R. Williams and his group in this country recognied the chemical structure of thiamine and then developed the first method for its synthesis.

Thiamine : Its Properties

It is white crystalline/substance soluble in water. It has nutlike, salty taste and a characteristic odour which in recognizable in the smell of yeast.

It is stable in acid but is rapidly destroyed in neutral or alkaline solution. It is not easily oxidized dry and can be heated in acid solution for half an hour at 120°C. in absence of loss but is destroyed by sulphite — a reality which would have some slight significance in fruit bottling but for the additional reality that fruit contains very little vitamin B_1.

It is also radically destroyed in the dehydration of meats and vegetables.

Commercially the vitamin is sold in the form of thiamine hydrocholoride.

Thiamine : Its function

It is vital to the body as one of the keys in the release of energy in the period of metabolism of carbohydrates where, as thiamine pyrophosphate, it functions in the enzyme system. Carbohydrate metabolism may proceed to the formation of pyruvic acid, an intermediate compound; thiamine is essential for the breakdown of pyruvic acid. It is believed that the enzyme systems containing the vitamins of the B-complex family, along with other vitamins and specific minerals, work cooperatively in the utilization and storage of energy and in the normal functioning of body processes that include growth and body maintenance.

Experiments have concluded that thiamine also may have a certain role in the functioning of peripheral nerves.

Thiamine appears to have severd indirect functions in the body since of its role in energy metabolism. The maintenance of the appetite, the tone of the muscles, and a healthy mental attitude are all related to thiamine nutrition.

One of the earliest sign of a thiamine deficiency, in man is the loss of appetite or *anorexia.* In numerou human studies involving thiamine deprivation anorexia was observed as early as the fourth day and became so severe as the deficiency progressed that the subjects had to force themselves to eat food. With the restoration of the vitamin to the diet, however, the loss of appetite disappeared very fact. Indirectly thiamine will eveourage growth since the stimulation of the appetite consequences in an increased consumption of food.

Constipation and tenderness of the muscles of the leg calf are two symptoms commonly found in experimental thiamine deficiencies. The sluggishness of the intestinal muscles concerned to constipation may be a since of inadequate thiamine in the diet. Soon after the vitamin is restored to the food intake after periods of thiamine deprivation, subjects exhibited relief from both of these symptoms.

Thiamine is usually called the *morale* vitamin since of its ability to restore a healthy mental state in man following periods of experimental deprivation. When human subject are fed restricted quantities of thiamine in their diets, changes in the emotional stability of normal subjects have been observed. Williams observed that his subjects on restricted intakes of thiamine became irritable, quarrelsome, and moody; they failed to cooperate; and at times they suffered periods of mental depression. Supposing that thiamine is not advantageou in the treatment of mental disorders, it has been observed that dietary restriction of the vitamin in mental patients appear to intensify the psychiatric symptoms of their mental state. The interest, ambition, attention, and sociability of these patients have been observed to be definitely affected by sub-optimal intakes of the vitamin as exhibited by a decrease in talking and a diplaying of body movement.

The Impression of a Thiamine Deficiency

In Animals. All species of animals require thiamine for their normal energy metabolism. Supposing that the ruminants (cattle and sheep) can synthesize substantial quantities of the vitamin in their rumens, other animals must attain a supply of thiamine in their daily food ration so as to meet their requirements.

Irrespective of*species, the symptoms of a thiamine deficienency in animals are quite similar. It is believed that the symptoms are related to the disturbances in carbohydrate metabolism accompanying thiamine deficiency, supposing that the precise relationship is not known as yet. The albino rat, which may be utilised in the classroom to demonstrate the impression of this vitamin deficiency, develops anorexia and immediately begins to lose weight when fed on a ration low on thiamine. As the deficiency progress, the animal develops a spastic gait since of a partial paralysis caused by affectation of the nerves; exhibits signs of a slowing of the heartbeat; regains its equilibrium with much difficulty after being rotated; and suffers from convulsions. The condition is known as polyneuritis. The recovery of these thiamine-deficient rats within eight hours after receiving supplements of the vitamin (Fig. 1)· has been observed.

In Man. Beriberi consequences from a severe deficiency of thiamine. Supposing that beriberi occurs very rarely in the Western countries, it is a general disease in the countries of the world where polished rice is the main item of the diet. The period from 1954 to 1958, beriberi ranked fourth as the reason of death among the people of the Philippine Islands.

The three common symptoms of beriberi are polyneuritis, ether edema or emaciation of the body tissues, and disturbances of the heart. The polyneuritis initiates with a soreness in the muscles of the legs and is immediately followed by a decrease in the reflex of the knee and a drop in the muscles that support the toes and foot. This latter symptom of the disorder accounts for the high-stepping gait observed in beriberi patients (Fig. 2). As the disorder advances, however, the arms and other parts of the body become affected following a degeneration and dearth of coordination of the body muscles.

When edema occurs in beriberi, the body tissues become swollen due to of the accumulation of fluid in the body cells(Fig. 3). This condition is known as *wet beriberi* and is noticed first in the legs and thighs of the patient. Occasionally a reverse condition wherein edema does not occur but readily emancipation exists; this is diagnosed as *dry beriberi.* Inasmuch as the beriberi sufferer breathes with very difficulty and his heart beats quickly and becomes enlarged (Fig 4), death in beriberi is since of these cardiac impairments. Beriberi heart disease has been found among some elderly hospitalized persons.

Cases of mild thiamine deficiency are quite general throughout the world. R.R. Williams (11) has written "... there are scores of millions, most likely hundreds of millions, in Asia, who are suffering from mild beriberi and have done so for years and they still do not know they have it..." In U.S.A., cases of extreme deprivation of the vitamin are rare, but sub-optimal intakes of thiamine have been reported to exist among people of all age groups as shown by nutrition surveys from dissimilar regions of the United States. From human studies of the impression of mild thiamine deprivation in the diet, one can recognize some of the typical symptoms that accompany subclinical cases. Fatigue, anorexia, constipation, laboured breathing, lassitude (weariness), nausea, tenderness of the leg calf, depression, changes in the electro-cardiogram measurement, low excretion of thiamine in the urine, and an increase of pyruvic acid in the blood are the symptoms which accompany a thiamine deficiency. Supposing that some of these symptoms are particular for thiamine, others may consequence from other nutrient deficiencies.

The Finding of the Nutritional Nature of Beriberi

The beri beri for a prolong time was concerned to the

orient privious it became famous to the Western world, so long and so widely, in fact, that the origin of the name is uncertain. Some trace it to one language in conformity with which it means a great weakness; others to the word for sheep (in a dissimilar language) since the first outward sign is usually a stiffness of the ankles offering the victim a "sheep-like" gait.

In formal terminology it has been called a *multiple peripheral neuritis,* a disease of the nerves of both feet. In severe cases it may consequence in paralysis of the legs, and sometimes eventually in heart failure. Sometimes the heart exhibits injury first.

In 1878-1883, the total strength of the Japanese navy was about 5000 men, and of these 1000 to 2000 each year were at some time on sick-list with beriberi. The enormous morbidity of 20 to 40 per cent from one disease led to an investigation by Takaki, a high medical officer of the Japanese navy, who worked upon the theory that the fault was most probably to be found in the food since climate seemed to have no ascendancy and the sanitary conditions on the Japanese ships were as good as those in the European navies which were not troubled with the disease. A Japanese naval vessel with 276 men on a 9 months' cruise from Japan to New Zealand, Valparaiso, and Honolulu had 169 cases of beriberi, with 25 deaths. On another vessel with a likewise crew, sent over the same route, but with a ration in which the rice was decreased, the barley increased, and vegetables, meat, and condensed milk added, only 14 men had beriberi and each of these had failed to eat his full allowance of the new foods. With this experiment, Takaki secured the adoption of a new ration for the entire Japanese navy with the consequence that the number of cases of beriberi immediately became practically negligible.

From his government, Takaki received generous value for this really great achievement; and yet it failed to accomplish what might appropriately have been expected of it. Opinion continued to be divided, as it had been before, as to whether beriberi were a nutritional or an infectious disease, with the majority of the medical men of the Orient (where the disease continued to be common) inclining to the theory of an infection.

Takaki, as we now know, was right in his belief that the disease was essentially nutritional and he was quite explicit in saying so, but his work did not carry conviction as he offered no sufficient theoretical analysis of it. He discussed his improvement of the ration only regarding protein, which others rightly considered as unconvincing. It was a time of great activity in the sanitary applications of the then new science of bacteriology. Sanitary improvements had obviously been made in the Japanese navy during the same years in which Takaki was succeeding in obtaining the ration changed, and the sanitary improvements were usually given credit for the eradication of the disease since the germ theory seemed sufficient even with the germ unidentified, whereas Takaki offered no satisfactory scientific analysis of what we he had accomplished.

Eijkman, a Dutch physician in 1897, working in the East Indies, published observations upon "an illness of fowls similar to beriberi." He had noticed that fowls living in the bare yard of a prison hospital and subsisting all but exclusively upon the left-over rice from the prisoners' table, usually exhibited stiffness and weakness of the feet and legs very suggetive of that of his beriberi patients. By systematic experimentation he found that this disease could continuously be induced in fowls by confining them strictly to a polished-rice diet. This is now often considered as the first clear-cut case of experimental production of a dietary deficiency disease. At the time, however, Eijkman's presentation and discussion of his findings was so strongly coloured by what English investigators afterward called "the pharmacological bias" that the bearing of his work upon normal nutrition was not made explicit. Reading Eijkman's early work in the light of subsequent knowledge, we may consider it as marking the beginning of our experimental knowledge of dietary deficiency diseases; but it was not until some years later, and until the experiments had been enxtended by Grijns (1901), that the consequences of this work were stated regarding a conspicuously nutritional concept.

Meanwhile British and American workers in the Orient were also starting to provide systematic attention to the nutritional aspects of beriberi.

For intance, when American Army officers took over the administration of the Bilibid prison in Manila they sought to improve the treatment of the prisoners by furnishing them a better commercial grade of rice-plump, pure-white rice of highest market grade instead of the "low-grade" irregularly shaped brownis rice previously purchased. Just thereafter, the proportion of cases of beriberi in this prison population started to increase; but for some time this was considerable as an epidemic to be fought by sanitary measures. After many months during which everything feasible in the source of sanitation had been done, but in absence of conquering the epidemic of beriberi, attention was turned to the "nutrition hypothesis." Dietary changes corresponding to those which Takaki had made in the Japanese navy were tried; and the beriberi was then brought under control.

Almost simultaneously. Fletcher, by experimenting with the diet in a hospital for the insane, exhibited that when 28 ounces of rice were fed daily with only small amounts of other food, the utilize of white polished, or of brown unpolished, rice was alone adequated to condition the occurrence or non-occurrence of beriberi.

In 1909, convinced that beriberi was concerned to diet, the United States Army Medical Commission in the Philippines initiated changes in the rations of the Philippine Scouts and in 1911 Champberlain was able to announce the eradication of beriberi from these troops by the improvement of their ration. In 1908-1909, when beriberi was at its worst among the scouts, the ration comprised essentially 12 ounces of beef, 8 ounces of white flour, 8 ounces of potatoes and 20 ounces of rice (ordinarily polished). The modify in the ration, as ultimately decided upon after some months of experimentation, consisted in giving, in place of the 20 ounces of polished rice, 16 ounces of unpolished rice and 1.6 ounces of dried beans. Experiments, made primarily upon fowls, had shown that while beef has some influence in preventing beriberi, an equal weight of beans, peas, or peanuts in much more efficacious. In conformity with chamberlain the disease had practically disappeared due to of adding the beans to the ration, privious the substitution of unpolished for polished rice had been completed. He held, "that the consumption of beans to the daily amount of 6 ounces would, unaided, have prevented a recurrence of beriberi but it would naturally be problem to ensure that all the men ate their share of this article over long periods, and it is therefore much safer that the bigest component of the diet, the rice, should be of the unpolished variety and by itself enough to prohibit neuritis." Many other investigations gave similar results.

Repeated demonstrations of a close connection between diet and the occurrence or prevention of beriberi naturally gave a great impetus to experiments designed to find, and to identify chemically, the precise materials which has the property of preventing and curing the neuritis which is the outstanding feature of this disease.

The Search for the Antinenritic Materials

Hulshoff-Pol in the beginning of 1902 showed that the antineuritic property could be demonstrated not only with the effective foods themselves, but also with extracts made from them, and even after clarification of similar extracts with fundamental lead acetate and later removal of the lead.

Fraser and Stanton, working in the Malay States, and Chamberlain, Vedder, and Williams in the Philippines, together exhibited the antineuritic constituent of foods to be an organic materials more stable in acid than in alkaline solution, perhaps a nitrogen compound, but not similar to any known amino acid or alkaloid. Funk studied the materials further and named it *vitamin* (e). Here the introduction of the name of the materials followed by about 10 years its clear apprehension as a definite chemical concept, and preceded by about 15 years its isolation by Jansen and Donath in 1926, while nearly another last 10 years of research was invested by Williams in working out a method of separating the pure materials in enough quantity to permit of the setting up of its molecular structure and full chemical nature.

Thus Williams, at great personal sacrifice of his "spare" time for a quarter century, achieved success in identifying the materials and formulating it available at tremendously low cost; and now has the satisfaction of seeing the saving of thousands of human lives through the utilize of the consequences of his work.

The Antineuritic Vitamin : Its Chemical Nature

Larger scale production was made feasible by the method of Williams, Waterman, and Keresztesy (1934), and ample authority soon endorsed and synthesis confirmed, the structural formula proposed by Williams.

The molecule contains a substituted pyrimidine nucleus and a substituted thiazole nucleus conneited by a $-CH_2-$ group. This was the first time that the thiazole nucleus had been found in a naturally-occurring substance.

The study of the chemical nature of this vitamin was greatly advanced by Williams' discovery that sulfite at pH 5.0 reasons a quantitative division of the molecule nbt two man fractions. The one which proved to be a thiazole derivative was detected and synthesized by Clarke and Gurin (1935); and Williams and Cline (1936) completed the synthesis and structural identification of the vitamin whose chemical nature Williams had been consistently investigating for about twenty-five years. Williams and Spies (1938) provide full authority for the structure, comprising an account of the synthetic work.

Thiamine as the Modern Name

After the molecular structure had been established, the name thiamine (thiamine chloride, thiamine chloride hydrochloride) was chosen as the official name of the materials by the Federal Food and Drug Administration, the American Medical Association, and the United States Pharmacopeia. It has also been adopted or supported by American scientific societies generally. The term vitamin B_1 continues to be utilitsed frequently synonymously with thiamine. The additional resemble, aneurin, sometimes encountered in European literature; but its utilize in some countries is discouraged.

Suggested Permission for Thiamine

The human requare for thiamine is based on the energy requirement of an individual. The 1958 revision of the Recommended Dietary Allowances proposed a thiamine value of 0.5 mg per 1000 Cal. In the 1963 revision, however, the suggested allowance for thiamine has been lowered to 0.4 mg per 1000 Cal.

College men and young male adults, 18 to 35 years of age, requare 1.2 mg of thiamine each day. A slightly lower allowance, 0.8 mg per day, has been suggested for the college and young adult woman of the same age group. In the stress periods of life (growth, pregnancy, and lactation) the thiamine requirement is increased. As one grows older, the requare for the vitamin decreases with the decrease in caloric need. However, new experimental findings connote that the minimum requirement for thiamine may be greater for the older person than for the younger one.

Thiamine : Its Sources

All the foods of the diet with the exception of fats, oils, and refined sugars contain some thiamine. However, there are no foods especially high in thiamine content as there are for ascorbic acid and vitamin A.

Although brewer's yeast and wheat germ head the list of the thiamine comprising foods (Table *7.1)*, they are not very potent in the typical food intake since relatively few people eat either one of these in large amounts. The best ways of the vitamin in the average diet are pork products, legumes, and whole-grain and enriched cereals. Nuts vary in their thiamine content but when they are utilised in large amounts, prove to be rich contributors of the vitamin. The thiamine in the egg is limited to the yolk. A quart of milk each day will contribute a good share of the thiamine need.

Table : Thiamine Values of Some Typical Foods

Food mg/100 gm	*Portion*	*mg/Serving*[2]	
Brewer's yeast	15.61	1 tbsp	1.25
Wheat germ	2.01	1 c	1.36
Pork chop, cooked	0.96	3½ oz	0.63
Brazil nuts, shelled	0.96	1 c	1.34
Pecan nuts, shelled	0.86	1 c	0.93
Ham, smoked, cooked	0.47	3 oz	0.40
Bran flakes, 40 per cent	0.40	1 oz	0.11
Green peas, fresh, cooked	0.28	2/3 c	0.30
Bread, whole-wheat	0.26	1 sl	0.06
Bread, white, enriched	0.25	1 sl	0.06
Lima beans, fresh, cooked	0.18	2/3 c	0.20
Peanut butter	0.13	1 tbsp	0.02
Rice, white, parboiled, cooked	0.11	1 c	0.19
Beef, ground, cooked	0.09	3 oz	0.08
Egg	0.09	1 med	0.05
Orange juice, frozen	0.09	1 c	0.21
Milk, whole	0.03	1 c	0.08
Sugar, white	0	1 tsp	0
Butter	0	1 tbsp	0

The utilize of enriched cereal grains has bolstered the thiamine value of the American diet. Data from the U.S. Department of Agriculture reveal that the national food supply contributed about 12 per cent more of the vitamin in 1962 than it did for the period prior to the adoption of cereal enrichment (1935-1939).

Thiamine Losses during Food Processing and Preparation

Vegetables : The quantity of thiamine retained in a vegetable during freezing or canning is explicitly concerned to the nature of the food used. Peas have displayed no significant loss of the vitamin during the freezing procedures, but spinach and asparagus have been reported to lose about 51 per cent and 28 per cent, respectively. In a study to condition the influence of canning procedures on the vitamin retention of vegetables, it was found that the thiamine retention was good in tomato juice, only far in asparagus, and comparatively poor in snapbeans, Lima beans, corn, peas, and spinach.

Under ordinary ssituations of home cooking, most of the thiamine lost in a food is due to the leaching of the vitamin into the cooking water readily than to heat destruction. This reality was observed when twelve different vegetables were processed in family-size amounts by four dissimilar cooking methods. The retention of thiamine in the vegetables averaged 91.0 per cent (waterless cooker), 84.1 per cent (tightly covered saucepan with ½ cup water), 78.9 per cent (pressure cooker), and 62.6 per cent (tightly covered saucepan with water to cover). In view of this report, it appears that the thiamine content of a vegetable can be conserved if it is cooked in the least amount of water compatible with good palatability.

Since, thiamine is highly soluble in water, the liquid fraction of a canned vegetable as well as the soaking or cooking water from a vegetable will contain an enough amount of the vitamin lost from the vegetable solid during processing or food preparation. To use all the thiamine available in the foods of the diet, this liquid fraction should be used and not rejected.

Meats **:** Because pork and its products are excellent ways of thiamine in the American diets, it is important to know the influence that cooking methods have on the retention of the vitamin in these foods.

Studies have revealed the retention of the vitamin in fresh hams and pork loins, fresh-ground pork, as well as in cured hams privious and after cooking by standardized procedures. Supposing that the thiamine content of dissimilar cuts from the same pork loin or ham may vary significantly, the retention of the nutrient after cooking is quite uniform. Pork loins and fresh hams have been shown to retain about 70 per cent of their thiamine after roasting and broiling and about 50 per cent after braising. When the vitamin content of the droppings from these pork products is added to the thiamine value of the cooked meat, there is about 70 per cent retention for all three modes of preparation. As much as 85 per cent of the vitamin was retained in ground pork patties when they were cooked by pan-broiling, oven-broiling, or electronic cookery. In the curing of ham, all but three-fourths of the original thiamine of the fresh meat has been reported to the retained. Fried cured ham has a higher thiamine content than cured ham that has been roasted, braised, or broiled. Regardless of the processing procedure or the method of cookery, it seems that cooked pork products are worth-while contributors of thiamine to the diets.

Thiamine : Its Normal Nutrition

While previous experiments of Lunin and of Pekelharing may now be believed to have dealt with what have since come to be known as vitamins, it was the work of Hopkins which first made the concept effectively exblicit to most students of normal nutrition. His alcohol extracts of dried milk and of specific dried vegetables were clearly exhibited to contain some materials or substances essential to growth. We now know that these extracts contained both fat-soluble and water-soluble nutritional factors, and that one of the latter was the materials now named thiamine.

Thus the longest-known functions of thiamine are the promotion of development and the prevention of polyneuritis.

Thiamine also has a potent function related to appetite. While all or most nutritional deficiencies tend to consequence in steady decline of appetite, thiamine deficiency reasons (after a short period in which the body is presumably using up its reserve) a characterically more abrupt loss of appetite than does any other known vitamin deficiency. Also, thiamine supports in the recovery of appetite in a more prompt and certain way than does any other known nutrient.

This vitamin formulates the food more appetizing and has an influence upon the appetite as a physiological function. For even the feeding of the thiamine separately reasons the animal to come back with appetite to the food which it had discarded.

Such observations endorsed the unqualified statement that this vitamin stimulates appetite. More recently there has been a tendency from two points of view to qualify this statement. From the standpoint of what is "permissible to claim" it has been proposed to limit the claim for thiamine's action on appetite to a statement that it promotes recovery of an appetite which has been lost due to of its lack. And in answer to the question whether stimulation of appetite is normally desirable it has been said that thiamine promotes recovery from a condition of subnormal appetite; and stabilizes the appetite at the normal level.

It may most likely be questioned whether the authority fully endorses these somewhat drastic limitations upon the hitherto recognised simple statement that one of the functions of this vitamin is to eveourage the appetite. Certainly some scientifically-trained people have approach they observed increases in already-normal appetites after taking thiamine. And it seems reasonable to suppose that a positive promotion of appetite is incpuded in the greater development which consequences from the higher levels of thiamine feeding among animals normal throughout.

Thiamine seems to function in the maintenance of the normal motility of the digestive tract. Both L.J. Harris and M.S. Rose wrote of a common lack of tone of the gastrointestinal tract following a shortage of this vitamin. Cowgill, experimenting with dogs, observed a diminution of vigor in the normal contractions of the stomach as an previous conequence of thiamine shortage than the loss of appetite. As the loss of normal motility may extend to the intestine, the repeatedly reported constipating influence of dietaries too largely composed of artificially refined food may perhaps be partly due to paucity of thiamine.

***Thiamine functions very importantly in the intermediary metabolism* :** An excellent succession of experiments by peters and his co-workers of Oxford has displayed that thiamine, presumably acting primarily in the form of its pyrophosphate (cocarboxylase), catalyzes the transformation of pyruvic acid and hence helps the carbohydrate metabolism through its intermediary stages. Thus thiamine carries the intermediary metabolism of carbohydrate through an essential step, whether on the source to combustion as fuel or to conversion into fat. This displays thiamine to be a materials of great significance to the working of the active cells in all the bodily systems.

Williams and Spies (1938) stressed this frequently and held that it affords a scientifically satisfactory analysis of the clinical reports of helpful consequences from thiamine therapy in a very wide variety of ills.

They also advised that the abnormal accumulation of pyruvic acid (and perhaps also other intermediates) in the body when thiamine intake is too low may be due to failure of appetite and of decline in gastro-intestinal motility and common bodily tone.

To what extent the aboundant clear "toning-up" property of thiamine is limited to "curing the consequences of a earlier lack of the same substance," and to what extent it stimulates an already normal appetite and bodily tone, is not yet clear.

Doubtless this is mainly since "B vitamin" has been found to include other things as well as thiamine.

Perhaps ten or more fairly independent substances may now (1951-2) be considered as usually recognised members of the B group of vitamins: (1) thiamine; (2) riboflavin; (3) niacin (niacinamide); (4) folic acid (pteroylglutamic acid or glutamates); (5) pantothenic acid; (6) pyridoxine (with pyridoxal and pyrdoxamine); (7) vitamin B_{12}; (8) choline; (9) biotin; (10) lyxoflavin; and from some viewpoints: inositol, para-aminobenzoic acid, and most likely other substances.

Thiamine in reproduction and lactation : There is substantial authority that "the vitamin-B complex" is importantly related in the nutritional support of successful pregnancy and lactation; but just how this accountability is shared between thiamine, riboflavin and other factors is still a difficult for further research.

Demonstration and Measurement of Thiamine in Foods

Methods Utilizing the Influence of Thiamine upon Growth of Mammalian Species : Rats served as subjects in the work of Hopkins, of Osborne and Mendel, and of McCollum which established the existence of this "new" factor as essential to normal nutrition, and the development methods for its determination have often made utilize of this species.

Such quntitative feeding tests can also easily have made to furnish striking demonstrations of both the prevention and cure of nutritional polyneuritis by thiamine and explicit evidence of its influence upon growth.

While the influence on the weight curve is usually the best basis for *measurement*, the most characteristic impression of lack of thiamine are failure of appetite and growth of polyneuritis. As the appetite fails there is also a growth of much the same situation as is yielded by starvation. Complete deprivation may even consequence in death in absence of the appearance of the characteristic nerve symptoms.

Other Methods of Thiamine Assay : Methods based on cure of thiamine-deficient rats or pigeons have been widely used, but more recently thiamine assays are usually made either by measuring its influence upon the activities of microorganisms (microbio-assays) or by *in vitro* methods.

Quantitative Expression of Thiamine Values

Whatever the method utilised in measuring thiamine values, good practice now requires that all quantitative testing include experiments with known amounts of the vitamin side-by-side with those upon the food or other material under investigation.

When properly conducted and interpreted similar quantitative experiments permit statement of the value found regarding the weight of the actual materials (thiamine, thiamine chloride, or thiamine chloride hydrocholoride).

When vitamin B_1 itself was available as a standard of comparison, units of activity were no longer requirement and it was feasible to express the potency of any food or pharmaceutical preparation in connection with the actual weight of vitamin naturally occurring in a provided weight of material, similar as, for example :

Food	*Vitamin* B_1 *in mg. oz.*
Wholemeals bread	0.06
Ox Liver	0.11
Fresh egg (whole)	0.04 (Ministry of Food, 1953)

Human Needs for Thiamine

It has been seen that thiamine helps carbohydrate (and doubtless also the glycerol radical of fat and the deaminized radicals of some of the amino acids) through the pyruvic acid stage of metabolism. Hence thiamine deficiency reasons in a relative accumulation of pyruvic acid in the body; and so, determinations of pyruvic acid in the body's tissues or fluids may be utilihed as authority as to whether there is a shortage of thiamine in the body at any provided time.

A number of investigators have made utilize of this as well as other criteria, compriing that of "thiamine clearance" (Melnick and Field, 1942) which may be considered as similar to the determination of percentage recovery in urine of test doses of ascorbic acid in the studies of vitamin C requirement.

In 1942 Melnick stressed the purpose nature of his thiamine balance studies as authority of requirements; and considered that from this authority the thiamine needs of the human adult could be estimated to be 350 micrograms (millionths of a gram) per 1000 Calories of food. To give a margin of safety and to cover individual differences he recommended 500 micrograms per 1000 Calories.

In 1947, Williams, Mason, and Wilder formulating use both of purpose chemical methods such as as Melnick used and the more subjective neurological tests and observations of preclinical symptoms in subjects living for long periods on thiamine-low diets, concluded that the daily allowance of 0.6 mg. (600 micrograms) per 1000 Calories, as then recommended by the Food and Nutrition Board of the National Research Council, is "none too high."

Melnick in 1944] in a subsequent paper concludes, from a consideration of the influences of various factors as well as individual differences, that the margin of insurance given by the allowance of 0.6 mg. of thiamine per 1000 Calories of food is to be considered as desirable.

In the 1948 revised Recommended Allowances, the thiamine is often set at 0.5 mg. per 1000 calories; but in pregnancy is increased in greater proportion than the energy allowance, while for men and boys with very high energy needs the thiamine allowance is not increased very in proportion to the caloric intake.

It must be kept in mind that the Recommended Allowances of the National Research Council are intended to represent intakes capable of supporting optimal nutritional well-being readily than as minimal needs. Thus we have, in utilize by the Federal Government at the same time, but for dissimilar objectives, an estimated minimum requare of 1000 mcg., and a recommended allowance of 1500 mcg., per day for in average-sized man of similar activity as to metabolize 3000 Calories per day.

The Recommended Allowances are : Men (70Kg.), sedentary, 1.2 mg.; physically active, 1.5 mg.; doing heavy work, 1.8 mg.; women (56 Kg.), sedentary, 1.0 mg.; moderately active, 1.2 mg.; very active, 1.5 mg.; pregnancy, latter half, 1.5 mg.; lactation, 1.5 mg.; children under 1 year, 0.4 mg.; 1-3 years, 0.6 mg.; 4-6 years 0.8 mg.; 7-9 years, 1.0 mg.; 10-12 years, 1.2 mg.; Girls, 13-15 years, 1.3 mg.; 16-20 years, 1-2 mg.; Boys, 13-15 years, 1.5 mg.; 16-20 years, 1.7 mg. of thiamine per day.

Thiamine In Typical Foods And In The American Diet

Thiamine is extensively distributed in substantial amounts

in foods of both plant and animal origin, except like have been artificially refined or otherwise denatured.

Plants synthesize thiamine and while it occurs in the dissimilar plant organs in their natural state, there is a common tendency toward a higher concentration of thiamine in the mature dry seed than in other parts of the plant.

Unmilled cereal grains are richer in thiamine than the diet as a all requires to be. Thus we saw more than that the liberal recommended dietary allowance of the National Research Council is 0.5 mg. of thiamine per 1000 Calories of food; but all wheat, all barley, and oats average about 1.6 mg. per 1000 Calories, and brown rice contains about 0.8 mg. per 1000 Calories. The major part of this thiamine is in or near the embryo (germ) and outer layer of the grain. Hence similar *milled products* as white flour, white rice, and degerminated corn (maize) meal or hominy, are relatively poor in this vitamin, as well as in others and in mineral elements. While the quantitative relationships differ with the grains and with the milling processes used, one in common can aspire that the highly refined grain product will contain only about one to two tenths as much thiamine as the original all grain, — or not nearly sufficient thiamine to give substantially for the metabolism of the carbohydrate carried by these foods. Thus diets made up primarily of refined grain products simultaneous refined sugars and fats hence enriched and a large proportion of breakfast tend to be undesirably low in thiamine. Fortification of similar products (legalistically designated as enrichment in the case of breadstuffs, and *restoration* in breakfast cereals) by additions of thiamine with or in absence of other nutrients came to be supported by a number of food chemists and to receive governmental approval and encouragement about 1940. By 1944 much the larger part of the flour and bread sold in the United States was enriched, and some of the States had made similar enrichment mandatory. With breadstuffs cereal hence enriched and a large proportion of breakfast restored to approximately all grain levels, the thiamine content of the American diet was (and is) greatly improved. The average thiamine intake of the people of the United States is hence brought to a level about 30 per cent more than the liberal recommendation of the National Research Council, and the distribution of our national consumption of thiamine is perhaps more equitable than of most other nutrients since in common more bread is consumed by low-income families. But, of course, to ensure this benefit to those who require it most, it is potent that enrichment of white flour and bread be made as nearly global as feasible.

Mature dry legumes rank with the whole-grain cereals in thiamine content. In this respect as in some others, there is overlapping between these two groups of foods, while within each group there is considerable variation among species and even among cultural sorts.

Among foods of animal origin the thiamine content is relatively low in milk, higher in eggs and lean meats, and among the meats it is especially high in pork muscle, while liver and kidney reveal an amount intermediate between lean pork muscle and the other muscle meats.

The other vegetables and the fruits vary still more widely, partly since of particular differences among plants, partly since vegetables include dissimilar organs of plants, and partly since both fruits and vegetables are marketed sometimes fresh with very high natural water content and sometimes dried.

Need of the B Vitamins

As vitamin B_1 is concerned to with carbohydrate metabolism, any conditions that include a higher calorific intake, e.g heavy manual work, climbing, Article exploration, obviously invite for higher intake.

In those diseases, too, in which the rate of metabolism is increased, more vitamin B_1 is needed.

Riboflavin and nicotinic acid also participate with vitamin B_1 in energy relations and the amounts required are concerned to the amount of vitamin B_1 In foods, these vitamins of the B complex occur together and a shortage of one usually implies a shortage of the others. It is only when the staple cereal happens to be grossly short of a particular factor as, for example polished rice and vitamin B_1 or maize and nicotinic acid, which well-defined deficiency diseases like beriberi and pellagra occur. In countries where mixed diets are customary, deficiencies are generally less severe but of a mixed type. For instance, a mild shortage of riboflavin will produce, in some individuals, roughness and redness of the skin over the nose and some cracking of the lips, readily cured by the consumption of extra quantities of one or two foods rich in riboflavin like eggs or liver.

Liveurse a shortage of nicotinic acid, not sufficient to reason pellagra, may produce digestive disturbances, a reddening of the skin of the face and hands and feelings of depression.

Nicotinic acid is related in health y people with the control of movements of precision and if its supply is limited subjects may find themselves unusually clumsy.

Different responses of dissimilar individuals to diets deficient in the B vitamins was puzzling until, as a consequence of a long succession of observations, it was learnt that these factors are also synthesized indorse the body of man and the higher animals though not by it. The intestinal tract comprises warm semi fluid material rich in nutrients and is, as we know, a medium in which there is usually an expensive bacterial flora. These organisms are not disease producing. They merely live in and with man. Their power of synthesizing vitamins is a relatively recent discovery. They appear to specialize in the factors of the B group but some can also synthesize vitamin K.

Ruminants, whose large rumen in the first part of the digestive tract is the site of vitamin formation, obtain so much riboflavin and nicotinic acid from bacterial activity that they fail to create reigns of deficiency even on diets lacking these vitamins. But in man the bacteria carry on their synthetic activities much further down the digestive tract and, supposing that undoubtedly some of the B vitamins fielded in this source are competent for human use, sufficient of each is not formed to satisfy all human requirements.

However, it is not problem to see that in conditions like diarrhea in which the contents of the alimentary tract move along quickly, less than the average amounts may be absorbed and symptoms of deficiency may occur if the determine is not checked. The recent utilize of some of the new drugs, sulphonamides and antibiotics has also divulged that these, too, assume the activities of the vitamin-B-producing bacteria. The drugs prohibit not only the development of disease-producing organisms but also the synthesis of the B vitamins in the intestine. This interference with production may be great sufficient to reason true secondary deficiencies in patients who may not have been very well supplied with vitamins before. With signs of peripheral neuritis cases have been observed, enlarged heart and swelling of the limbs, indistinguishable from true beriberi. Other patients create a sore mouth and red scaling of the skin recalling deficiencies of nicotinic acid and riboflavin.

In the treatment of these secondary deficiencies very large doses have to be given — a hundred or more times the amount needed for normal health.

6

Riboflavin : Its Nutrition

Historical Backdrop

This was the first of the heat-stable factors to be distinguished. It is yellow, fluorescent pigment, which had been invented in milk in the 19th century, but this was privious we had any knowledge of vitamins and it was only when, in 1933, its property of encouraging development in rats was invented that it became involved among the other vitamins. It occurs eggs, liver, yeast and cereals. It was synthesized in 1933 and is still sometimes referred to as vitamin B_2 though this is obviously misleading now that we are aware of the multiple nature of what was originally called, vitamin B_2

In the beginning of 1920 some workers started questioning the belief that *water-soluble* B was a single nutritive factor. Studies connoted that after a food containing this dietary necessary had been autoclaved, if still had the ability to foster growth.

In animals even though its antineutritic powers were destroyed. This observation connoted that the water-soluble, vitamins B was not a single factor but readily a mixture of factors, one that was destroyed by heat and the other that heat did not affect. About 1928 the reality had definitely been established that there were at least two separate vitamins available in this complex, an antineuritic factor, Vitamin B_1, now called thiamine and a growth-promoting factor Vitamin G or B_2, now called riboflavin.

The first steps in the isolation of riboflavin taken in Germany about 1930 when a yellowish-green fluorescent material, which stimulated development in animals, was isolated from egg white. Soon it was found that likewise fluorescent substances were present in milk, liver, yeast, muscle, and particular plant materials. The fluorescent substance all proved to belong to a chemical family called the *flavins* and so were named after the foodstuff from which they were isolated, like *ovoflavin* from egg and *lactoflavin* from milk. Later chemical studies displayed that all of these fluorescent substance were identical with one another and the name, riboflavin, was the common term adopted to recognise them. Riboflavin has been yielded synthetically since 1935.

Riboflavin : Its Properties

Riboflavin occurs in yellow-orange crystals (Fig. 1), which are only slightly soluble in water; solutions of the vitamin have a yellowish-green flourescence. The availability of acids, air, or heat have no destructive influence upon the vitamin, but riboflavin is readily destroyed by alkalies and light.

Riboflavin : Its Function

Like thiamine, riboflavin serves as a coenzyme in several of the enzyme systems of the body that are involved in the metabolism of the energy nutrients.

Occurrence of Riboflavin

Riboflavin (sometimes free, sometimes in combination with phospohoric acid, often in combination with both phosphoric acid and protein) is very much disturbed in both plant and animals tissues. The so-called respiratory enzyme

(Warburg's yellow enzyme) is a combination of riboflavin phosphate with protein. Many similar combinations are now known. The working-out of their interrelationships as catalysts of the oxidation processes in the tissues is an active sphere of research.

The riboflavin-phosphoric acid-protein combination of typical yellow respiratory tissue enzyme was mentioned by Kekwick and Pedersen (1936), on the ground of sedimentation, diffusion, and electrophoresis measurements, as a chemical entity with a molecular weight of the order of about 80,000 and with one flavin group in similar a molecule.

In milk, the riboflavin exists partially as an analogous compound, but also very largely in easily diffusible form, evidently free from combination with any large protein molecule.

Whether riboflavin in food is in the free state, or in combination with phosphoric acid, or with phosphoric acid and protein, perhaps formulates little difference to its nutritive value. If we believe one view riboflavin, if free when it reaches the intestine, is combined with phosphoric acid in the course of its absorption through the intestinal wall.

Thuse statements of riboflavin values of foods usually do not distinguish as to whether or in what proportions, it exists in a free or a combined shape.

The Impression of a Riboflavin Deficiency

In Animals. As the rat was utilised as the experimental animal in the early studies of this vitamin there is available a wealth of information on the influence of a riboflavin deficiency in this species. Growth failure occurs even though the animal is ingesting a considerable quantity of food, is one of the *early* symptoms of Riboflavin deprivation in the rat. Soon changes in the condition' of the skin and eyes become apparent.

The eye lesions in a riboflavin deficient rat vary in severity and may range from *conjunctivitis* to the formation of a permanent cataract. Digestive and nervous disturbances are also observed as the deficiency becomes more pronounced, and death will occur if the vitamin is not restored to the animal's ration. All animals require riboflavin for normal development and body maintenance.

In Man : Riboflavin deficiency on man may be complicated by a deficiency of other vitamins of the B-complex family, and following this cause the clinical symptoms of human riboflavin deprivation are not as clear-cut as are those of thiamine. The first experimentally induced riboflavin deficiency in man was reported in 1938 by Sebrell and Butler. They found that the women subjects of their study developed *cheilosis*, a reddening of the mucous membranes of the lips, and *seborrhea.* All of these lesions disappeared when riboflavin was restored to the diet. In a more recent study of induced human *ariboflavinosis*, the same sort of skin lesions were found to develop in the male subjects maintained on suboptimal intakes of the vitamin for a period of 9 to 17 months Another lesions reported by these workers was scrotal dermatitis. The changes in the tissues of the eyes since of riboflavin deprivation in humans are not as particular as the lesions of the skin. It is believed that *conjunctivitis* as well as *lacrimation* (a watering of the eye) are the ocular lesions which can be attributed to human ariboflavinosis. The condition involving the *vascularization* of the *cornea* (where the capillaries invade the cornea tissue) has not been proven as yet to be a certain lesion of riboflavin insufficiency in man.

Factors that May Affect Riboflavin Metabolism

The excretion of riboflavin in the urine is a purposeful index to reflect the state of the metabolism of the vitamin in the animal organism. It is imagine that a marked increase in urinary riboflavin connotes a loss of the nutrient from the tissue whereas a decrease connotes a conservation of tissue of riboflavin. Recent studies with animals and man suggest that environmental, nutritional, and/or physiological factors may assume the normal metabolism of riboflavin.

The impression of exposure to sunlight (ultraviolet irradiation) for diverse periods of time on the riboflavin metabolism of the rat has been investigated at the Institute of Nutrition, USSR Academy of Medical Sciences. Moderate exposure of the animals to these rays clearly increased the capacity of the tissues to retain the vitamin, but there was a marked loss of tissue riboflavin during prolonged exposure. A enough supply of dietary riboflavin, however, increased the animal's tolerance to ultraviolet light.

The ascendancy of sleep, bed rest, work, diuresis, heat, acute starvation, and thiamine intake on the riboflavin exertion in man has been studied. Sleep retarded the exertion of the nutrient whereas the rate of exertion was enhancess in the subjects during enforced bed rest. There was a decreased excertion of riboflavin in small periods of hard physical work. Diuresis, induced by water, did not influence the excretion rate of the vitamin. Both heat stress and acute starvation increased the excretion of riboflavin in the subjects. There was an immediate increase in riboflavin excretion when thiamine was diminished from the diet but with a combined deficiency of thiamine and riboflavin, the increase in the loss of riboflavin was delayed.

Suggested Permission for Riboflavin

The requirement for riboflavin is related to the allowance for energy, as are those for thiamine and niacin. The daily riboflavin requare of college and young adult men, 18 to 35 years of age, has been suggested to be 1.7 mg whereas college and young adult women of the same age group have been assigned 1.3 mg of the vitamin each day. Middle-aged men, 35 to 55 years, requare 1.6 mg and older men, over 55 years, requare 1.3 mg of riboflavin each day. After the age of 35, the daily requirement for women has been proposed to be 1.2 mg per day. Here again, the requare for this vitamin is increased during growth, pregnancy, or lactation.

Riboflavin : Its Sources

Since the best food ways of riboflavin, brewer's yeast and

glandular meats, are not used to any great extent in an average diet, their value as dietary contributors is not very coniderable. However, since it is so extensively utilised as a dietary food, milk is perhaps the main way of vitamin in most diets. The ingestion of 3 cups of all milk per day contributes about 1.26 mg of the nutrient, which is about 74 per cent and 97 per cent, respectively, of the daily riboflavin requirement of college men and women over 18 years of age. Cheese, eggs, veal, beef, leafy green vegetables, and salmon are other foods that are good ways of the vitamin. Apart from, the utilize of all grain or enriched cereal products readily than those that are unenriched will bolster the amount of the vitamin ingested. If we believe statistics from the U.S. Department of Agriculture, the national food supply in 1962 contributed about 12 per cent more riboflavin than it did in the period (1935-1939) privious the adoption of cereal enrichment. Students usually wonder why a 100 gm portion of all wheat bread (0.12 mg) has less riboflavin than an equal amount of white enriched bread (0.21 mg). The answer to this is that more of the nutrient is added to the wheat flour duringthe enrichment process than was originally present in the all wheat grain.

Riboflavin Losses during Food Processing, Storage, and Cooking

The two factors in the processing, storage, or cooking of a food that reasons a loss of riboflavin are the leaching of the vitamin into the processing or cooking water and the destruction of the vitamin by light.

The loss of riboflavin during food preservation and preparation takes place often during the blanching (boiling for a short period prior to preservation by freezing or canning) or cooking process. It has been reported that peas lost 20 per cent of their riboflavin content while being blanched, but relatively no further loss of the vitamin occurred in the subsequent stages of freezing or canning. Losses of the vitamin following of varying quantities of cooking water have been observed. There appears to be about a 20 per cent loss of riboflavin from vegetables cooked in a small amount of water (one-half cup), and when the water is increased the retention of the vitamin is decreased. However, very little riboflavin is destroyed in the range of temperatures used in general cooking practices following the heat-stable nature of the vitamin. Supposing that the solubility of this vitamin is not as great as was found either for ascorbic acid or for thiamine, the liquid fraction from a processed food should be used with a view to reclaim that part of the vitamin that has leached out into the processing water.

The destruction of riboflavin by light is a serious difficult due to the reality that one of the important dietary ways of the vitamin, milk, is often exposed to it. When a bottle of milk is permitted to stand on the doorstep on a bright sunny day for a period of two hours, as much as 85 per cent of its riboflavin value may be destroyed. On the contrary, the exposure of a bottle of milk for one hour on a rainy, dark day may consequence in only about 10 per cent of the vitamin being lost. Most dairies recognize this difficult and have made a constructive effort to preserve the riboflavin quantity of this food product. Both paper cartons and amber-coloured bottles have been introduced to remove light destruction of milk riboflavin.

Riboflavin : Its Nutritional Chemistry

The journal of the American Medical Association has spoken of riboflavin as "essential to the defense powers of the organism"; and Pinkerton and Bessey (1939) working in the Harvard Department of Pathology have exhibited that the level of nutritional intake of riboflavin widely affects the susceptibility of experimental animals to rickettsial disease.

There is sufficient authority to point out that the amount of riboflavin in the food has a large ascendancy upon the level of health and efficiency of people living under ordinary environmental conditions. This is a reality of basic significance and should be held explicitly in mind when one meets differences of view in respect to the importance of diagnostic signs.

For, while there may be doubt as to the relations of riboflavin to disease, there is no doubt as to its great importance to health, efficiency, vigour, and resistance.

Concerning Level of Riboflavin Intake to Growth, Development, Aging and Longevity

The concerning of riboflavin to development is similar that through an important period of the growth of our knowledge of riboflavin values of foods, these were determined by means of development experiments. That the development rate (of animals receiving all that the requirement of other nutrients) is so directly dependent upon the riboflavin intake is easily intelligible when we remember that riboflavin is an very important constituent of all the active body tissues that have been examined for it, and that the riboflavin enzyme is known to be a fundamentally important part of the muscle tissue of which the body is so largely composed.

Since these riboflavin enzymes are so definitely built into the very important substances of muscle tissue, that tissues retain their riboflavin so tenaciously under dietary deprivation. And riboflavin deficiency does not reason so early and quick a loss of health and body weight as does an equal shortage of thiamine. And when, there is not a total deprivation but readily a suboptimal intake of riboflavin, there may be nothing which ordinary medical examination would detect as a deficiency disease. In the young, there may be only a slowness of development and growth which the ordinary examiner would consider within the normal range and would either ignore or attribute to inherent physiological variability.

The slowing of development which consequences from suboptimal intake of riboflavin is accompanied by a imilar retardation of common development. And if shortage of riboflavin on the food supply continues throughout the period of development we must expect that the individual's growth will remain below the level of his inherited or inborn potentiality.

It may be hoped that similar chronic paucity of riboflavin intake will lessen life, and that to a still greater extent the period of enjoyment of the individual's full adult capacity will be reduced by the early onset of senility.

Whether or not one considers riboflavin deficiency as a repeatedly occurring disease, it is clear that riboflavin is a potent factor in health, and in the building of already-normal health to higher levels.

Hence riboflavin is necessary to growth, and to normal nutrition at all ages. When the food is distinctly poor in riboflavin for any significant length of time, digestive disturbances, nervous depression (different from the symmetrical polyneuritis of thiamine deficiency though both may occur in the same person), general weakness and deterioration of tone, and poor conditions of the eyes and skin are not to develop; the incidence of infectious disease is likely to be increased, potentiality diminished, life shortened, and the prime of life reduced by the unduly early growth of the physical manifestations of old age.

In comparison to sufficient or minimal sufficient intake of riboflavin, a diet of higher riboflavin value tends to consequence in better development, higher adult vitality, greater freedom from disease at all ages, somewhat longer life, and (more significantly) a longer "prime of life", i.e., a longer section of the life cycle between the attainment of adult capacity and the onset of old age.

This extension of the "period of the prime" — a property not of riboflavin alone but of the superior nutritional status or internal environment in which riboflavin is explicitly one of the major factors — has been particularly emphasized by Simms as important both to the individual and the community since regarding human affairs it means longer life with a *smaller percentage of years of dependence.*

In controlled experiments extending through two generations of laboratory animals, it has appeared: (1) that riboflavin ranked with calcium and vitamin A as the great factors in the improvement of nutritional welbeing which resulted from an increase of the proportion of milk in already-adequate diet; and (2) that in experiments in which riboflavin was the sole significant variable, successive increments of intake resulted in successively increased advantages to nutritional well-being, up to levels above twice that of minimal adequacy.

Relation to the Health of the Mouth, Skin, and Eyes

In their first description of riboflavin deficiency as a human disease for which they coined the name *ariboflavinosis,* sebrell and Butler stressed chiefly the symptom which consists in an inflammation of the lips with cracking at the angles *(cheilosis).* Frequently accompanying symptoms are a par kind of glossitis (sore mouth with an abnormally smooth purplish-red tongue) and a dermatitis around the folds of the nostrils. correspondingly, dermatitis was a general symptom in the human subjects maintained by Horwitt, et al. (1949) for long periods on a low intake of riboflavin. In riboflavin-deficiency experiments with rats, the dermatitis frequently manifests itself in any of three ways: as "sour nose," or by loss of hair around the eyes, or on the shoulders, the latter two offering rise respectively to the familiar names "spectack-like" and "saddle-like"

Day and his coworkers proved that riboflavin deficiency deterioration not only of the skin but also of the eye (which by origin is a specially developed skin-spot). The correlation of the skin and eye conditions with each other and with cessation of development in the young, and the experimental conditions under which these three signs of deficiency are governed by riboflavin, were apparently developed in the independent succession of papers by Day and by Bessey with their respective coworkers. Their discovery of the sensitiveness of the eye to shortage of riboflavin, made by means of controlled experiments with laboratory animals, has since been strikingly confirmed in the clinical experience of Sydenstricker, Sebrell, Kruse, and Others:

Directy clinical authority that riboflavin-deficiency is not only a possibility but and actually occurring disease in the United States, is furnished by the work of Sebrell and Butler (1938), Sydenstriker, et al. (1939), Oden, Oden and Sebrel (1939), Kruse, Sydenstriker, and Sebrell (1940), Sydenstriker, Sebrell, Cleckley, and Kruse (1940), Kruse et al (1943), and by several other medical observers. Thus Spies and others have emphasized strongly the frequency of its occurrence in our South; and Aykroyd in a lecture at the U.S. Department of Agriculture in 1943 stated that riboflavin deficiency is the most frequent of the vitamin deficiencies in India, white Braun, et al. (1945) found riboflavin deficiency in 190 of 900 pregnant women examined in Palestine. It has also been reported as of frequently occurrence in many other countries Medical reports have stressed its occurrence both independently and in conjunction with pellagra. In the latter cases, the much more dramatic symptoms of the pellagra, along with the reality that the two diseases generelly both yield promptly to good diet, make it likely that a large proportion of the riboflavin deficiencies are overlooked. For the same causes it is perhops that the pathology of ariboflavinosis as hence far known constitutes the beginnings readily than a complete picture of the consequences of riboflavin deficiency.

Slit-lamp examination showed an abnormal vascularity technically designated as "a superficial vascularizing keratitis" with "congestion and proliferation of the limbic plexus" as the earliest and global symptom in a succession of 47 cases, all of which were shown to be riboflavin deficiencies by the readiness with which they were cured by riboflavin alone. Some of these people were hospital patients who had been on "pellagra-producing" diets but were saved from the pellagra symptoms as now comprehended by means of nicotinic acid. Others of the people in this succession of 47 were members of the hospital staff who considered themselves well nourished and who came under observation simply due to cracked lips, eyes unduly sensitive to light, dimness of vision, or "eyestrain." While the latter people had chosen their food at will from a clearly sufficient food supply, investigation revealed that: "Bad dietary habits with insufficient intake of milk, eggs, and green vegetables were prevalent in the entire group." This investigation adds strong evidence for the view that a large proportion of people require a dissimilar emphasis in their choice of food so as to get sufficient riboflavin to support them at the best level of health of which they are capable.

However, two other realities must be kept in mind in this connection. There is not yet an entire agreement of medical opinion as to how far the eye symptoms may be taken as authority of riboflavin deficiency. And, while unwise choice of food may be liable for much of the riboflavin deficiency that exists, poverty is also undoubtedly often a primary factor.

Relation to Incidence of Infectious Disease

It has now been definitely exhibited that the level of intake of riboflavin has a marked ascendancy upon a specific type of infectious disease in experimental animals; and it is correct that this is likewise true for man, and for an as yet undetermined number of other infections. A already we have noted the American Medical Association based its official "recognition" of riboflavin on the common basis of its function in promoting and maintaining the resistance of the tissues. And this view is, of course, strongly advolated by the growing significance of the importance of riboflavin as a constituent of tissue enzymes and in the promotion of a high common level of vitality and vigor. Thus there is a broad and strong ground for the common recognition of liberal riboflavin intake as a factor in resistance.

Problems of Interrelation of Riboflavin with Other Factors

The occurrence and importanice of riboflavin in dissimilar combinations and its feasible interrelations with analogous derivatives of thiamine and of nicotinic acid are subjects of active and complicated research.

Here it will be sufficient to call attention to the reality that the findings of many investigators show that compounds (we do not yet know how many) of thiamine, of riboflavin, of nicoitinic acid, and of other B-complex vitamins all function in the complex enzyme-coenzyme systems which catalyze the oxidation process in the body tissues, and that there may be interrelations in their action. It is partly through studies of the enzymic reactions *in vitro* and partly through very extensive feeding experiments under exceptionally proper laboratory control that one may as pire to have further light upon similar questions as : to what extent do the thiamine, the riboflavin, and the nicotinic acid compounds function independently; to what extent are any of their functions interlinked, so that shortage of one impairs the functioning of another; or do any two of them act in parallel to a large extent that an abundance of one may in some degree ameliorate the influence of a shortage of one of the others?

The discovery by Ellis and Zmachinsky (1937 and subsequent on experiments) that as high level of riboflavin intake increases the ability to withstand a deprivation of thiamine; may afford an affirmative answer to this last question; or may be explained upon the mass-action principle with the imagination of a general dissociation product inasmuch as thiamine and riboflavin both contain the pyrimidine nucleus.

Riboflavin has also been found to increase hemoglobin production in dogs which had been rendered anemic by bleeding and then fed standard basal diet.

Interesting also is the work of Stannus (1944) identifying riboflavin deficiency with "*pellagra sine pellagra*" and recording his findings as to the sequence in which dissimilar body tissues are attacked in riboflavin deficiency: which Stannus prefers to call *hyporiboflavinosis* readily than ariboflavinosis.

Quantitative Measurement and Evaluation of Riboflavin Values

At present the riboflavin values of foods are considered

as following the one materials and within the limits of formal nutrition at least, the nutritional availability, and efficacy of riboflavin is inevitably the same whether taken in the free state or in the form or forms of combination in which it largely occurs in plant and animal tissues.

Riboflavin is now so readily available in pure form that all quantitative work on riboflavin values of foods should be controlled, standardized, and interpreted by paralled operations with the pure substance, and the consequences are best expressed in connection with weight of actual riboflavin. Milligrams micrograms (mcg) of riboflavin per 100 grams of food are two scales of expression now usually considered as most: As in the case of thiamine, *in vitro* methods and microbioassays for the determination of riboflavin have been developed and are in common use.

In the meantime quantitatively controlled feeding methods in which side-by-side experimental animals receive in some cases graded amounts of riboflavin and in other cases graded amounts of the food under investigation, may be used as a "court of eventual resort."

In drawing the averages for the accompanying table and in compiling the riboflavin data in the Appendix, consequences received *in vitro*, by microbioassay, and by rat-feeding experiments have all been used.

Foods as a Sources of Riboflavin

Riboflavin is formed in the development of green plants whose leaves and stems seem to be richest in riboflavin when in their most succulent stage, despite the high water content.

Liver and kidneys, milk (whole or skimmed, fresh, canned, or dried), cheese, cream, eggs, lean meats, legumes, and green leaves and buds are typical foods of relatively high riboflavin content. We should not lean too heavily upon comparisons of reported riboflavin contents in foods of different sorts, since some of the methods of determination are less proper for some kinds of food than for others. Hence comparisons *among* meats, for instance, are presumably more proper than comparisons of miscellaneous vegetables *with* meats.

If we believe a current figures dark meat is much richer in riboflavin than light meat of poultry. Unlike the case of thiamine, riboflavin values of lamb, beef, and pork muscle do not differ considerably. On the other hand, liver and kidney contain at least ten times as much riboflavin as the. muscle meats, and are among the most concentrated food ways of this factor. Perhaps it is significant that eggs average higher in riboflavin than muscle meats; and cheese (of Cheddar and correponding types) higher than eggs. Among the vegetables and fruits, the deep-green leaves and green flower buds (broccoli) are outstanding, containing maintines as much riboflavin as do typical fruits and other-than-green vegetables. The reality that sweet corn with its high water content nevertheless has about the same riboflavin content as does mature dry corn illustrates our common aspiration of vegetables in the succulent stage as good sources.

Table : Riboflavin in Edible Portion of Some Typical Foods: Milligrams Per 100 Grams

	Range within which the normal average may be expected to be found
Almonds	0.5-0.8
Apple, raw	0.02-0.05
Beans, snap or string	0.10-0.15
Beef muscle	0.15-0.3
Broccoli	0.15-0.3
Cabbage	0.05-0.3
Cheese (Cheddar type)	0.05-1.0
Eggs	0.3-0.5
Kale	0.25-0.4
Liver	2.0-4.0
Milk	0.17-0.25

Peas, fresh green	0.1-0.3
Pork muscle	0.2-0.4
Potatoes	0.03-0.05
Sugar	none or negligible
Sweetpotatoes	0.05-0.1
Tomatoes	0.03-0.05
Turnips	0.05-0.10
Wheat, entire	0.10-0.22
Wheat germ	0.6-0.8

The dry, mature, entire grains of barley, corn, oats, and wheat show very corresponding amounts of riboflavin, only about one fourth as much as their thiamine contents.

These all grain cereals hence contain considerable but not large amounts of riboflavin. Hence the addition of riboflavin in the enrichment of white flour and bread, while bringing the riboflavin content to about "whole-grain-levels" does not make nearly so large a contribution toward meeting the consumer's requare in the case of riboflavin as in the case of thiamine.

Riboflavin Needs in Human Nutrition

The Food and Nutrition Board of the National Research Council in its 1948 Allowances, recommended as follows:

Table : Recommended Dietary Allowances of Riboflavin

Child under 1 year	0.6 mg.
Child, 1-8 years (12 kg.)	0.9 mg.
Child, 4-6 years (19 kg.)	1.2 mg.
Child, 7-9 years (26 kg.)	1.5 mg.
Child, 10-12 years (35 kg.)	1.8 mg.
Girl, 13-15 years (49 kg.)	2.0 mg.
Girl, 16-20 years (55 kg.)	1.8 mg.
Boy, 13-15 years (49 kg.)	2.0 mg.
Boy 16-20 years (64 kg.)	2.5 mg.
Woman in second half of pregnancy,	2.5 mg.
Woman in lactation	3.0 mg.

The discussion accompanying the Recommended Allowances is, in part, "Evidence for assessing the desirable allowance of riboflavin is still readily incomplete....Keys *et al.* (1945) observed no physical or functional deterioration in a group of active young men who for nearly six months were supplied with only 0.29 mg. riboflavin per 1000 calories; whereas, Sebrell, Butler, and Wooley (1941) found that in 3 to 8 months a group of sedentary women receiving 0.21 mg. riboflavin per 1000 Calories developed distinctive physical signs of deficiency, which were responsive to riboflavin therapy.... the young men, due to their greater food consumption, received a total of nearly 1 mg. of riboflavin daily whereas the sedentary women ingested but o.5 mg per day....

"Williams, *et al.* (1943) found no clear handicap in four female subjects maintained for 9 months on 13 mcg. of riboflavin per kg. per day. However, neither 13 nor 17 mcg. per kg. prevented a specific degree of depletion of tissue riboflavin...."

"It appears likely that very long experimental periods would be needed privious the resulte of milder deficiency states would become clear."

Thus the situation is that all the experiments with human beings have been for too short periods to show what the allowance would require to be to bring optimal well-being (so far as this factor is concerned) throughout the normal life cycle. The present Recommended Allowances are, in the judgment of the present writer, above minimal sufficient but less than completely optimal.

There are other cause also more over mentioned more than for liberal intakes of riboflavin in human nutrition.

Spies, et al. (1940) have found that riboflavin deficiency "seems to be the commonest of the clinical deficiency diseases among children of the Alabama area studied."

Sydenstricker considered riboflavin deficiency to be perhaps the most frequent of all vitamin deficiencies in the United States; and Akroyd approach this also true of India.

It has been found that dogs on diet deficient only in riboflavin exhibit no ill-effects for a comparatively long time (102 to 140 days), then undergo a sudden decline of collapse. Pure riboflavin provided by injection or added to the diet prevents the breakdown or cures it if provided in time.

It may be taken as strong evidence of the vitality of riboflavin to the internal environment, that the body tends to hold its riboflavin content tenaciously, both in cases of lack of intake of this vitamin, or of pellagra patients.

Urinary exertion hence becomes a means of judging the adequacy of the body's supply 6f riboflavin.

Feder, Lewis, and Alden (1944) found "a widespread incidence of riboflavin deficiency," as judged by measurements of urinary excretion of riboflavin, in the Southeastern area of the United States.

Stiebling and Phipard's dietary studies also display a marked prevalence of riboflavin-poor dietaries in the Southeastern states; and in low-cost dietaries of other regions as well.

The wide unexplained variation of riboflavin contents of normal meals found by Kerly (1944), for example, from a low of 0.28 to a high of 0.93 mg. in 14 "entire dinners" served in school canteens, gives added emphasis to the advice that dietary allowances of riboflavin may well be made more liberal than those which are usually accepted as standard.

The consumption of riboflavin in per capita the food of the people of the United States as officially estimated for 1943 and 1944 is practically the same as the Recommended Allowance figured to a per capita basis from the Census data of the age distribution of the population. Hence dietaries of the riboflavin- content indicated by the Recommended Allowances are not unduly problem of attainment, assuming average purchasing power and current knowledge of food values. But when decisions as to what foods to buy, and how much of each, are not guided by nutritional knowledge, the higher-income families with their larger per capita purchases of meats, eggs, milk, cheese, cream, and ice cream get more, and the lower-income families get less, than their *pro rata* shares of the riboflavin in the nation's food supply.

One objective of wise nutritional guidance is to raise the lower levels of riboflavin consumption in absence of essentially lowering the higher levels. In the United States this requirement not be difficult. It can most easily be bought about by increasing the production and consumption of milk and its products other than butter. This group of foods supplies at present levels about half the riboflavin of the typical American family dietary; and to enhance the utilize of food of this group by perhaps one third in the average lower-income family will give similar a family a better-balanced dietary, whatever be the shift in the food budget by which the purchase price is provided.

7

Niacin : Its Nutrition

The disease, *pellagra*, which is now known to be prevented by an sufficient body supply of niacin, has been known for centuries. Supposing that it was once an endemic disease caused by the intake of a diet referred to as the "Three M's" (meat—salt pork, whichi was all but all fat and no lean—maize, and molasses) and was found to occur in the southern part of the United States. Pellagra is still prevalent in particular sections of Italy, Spain, and Rumania where corn products are a staple part of the diet. The characteristic symptoms of pellagra involve the skin, the gastrointestinal tract, and the nervous system. Following its certain lesions, pellagra is known as the disease of the "Three D's" (dermatitis, diarrhea, and dementia). The characteristic dermatitis of pellagra, which is aggravated by sunlight and heat, is bilaterally symmetrical and occurs on the hands, feet, face, forearms, and other parts of the body (Fig. 1) The skin becomes red and the lesions itch and burn. When the mucous membranes of the gastro-intestinal tract become affected, the early symptoms comprise *glossitis*, anorexia, and abdominal pains. As the disorder becomes more severe, diarrhoa occurs. Insomnia, irritablility, fear, depression, and forgetfulness are the first neurological changes observed but as the disorder advances, dementia and paralysis may occur. If pellagra remains untreated, it may prove fatal.

Active work to condition the reason of pellagra started in 1914 when the disease was very prevalent in southern states of U.S.A. Supposing that it had been advised that pellagra was related to the intake of a poor diet, there were numerous people who believed it was an infectious disease. Goldberger and his coworkers, under the auspices of the U.S. Public Health Service, were the first to induce and then cure pellagra in healthy men by dietary means. Soon after this classic experiment it was found that brewer's yeast eaten in satifactory amounts would protect a human against goowing pellagra.

Mean while the discovery of a disorder in dogs known as *blacktongue*, which is analogous to human pellagra, was the symptom in the test animal that led to the eventual recognition of this dietary factor. At the University of Wisconsin in 1937, Elvehjem and his group isolated a compound known as nicotinamide from a liver concentrate and then demonstrated its influence and that of a related compound, nicotinic acid, in curing and preventing blacktongue in dogs. Supposing that the chemical nature of nicotinamide had been known since 1870, its biological value was not detected until this time. The first successful treatment of human pellagra with either nicotinic acid or nicotinamide was reported the same year. To get rid of confusion with the name of the tobacco extract, nicotine, the name of the vitamin, nicotinic acid, was changed to niacin immediately after its discovery. It was Cowgill at Yale who coined the name, niacin, from the letters of the three words—nicotinic acid, and vitamin.

Niacin : Its Properties

Niacin is found as colourless needles having a very bitter taste. It is soluble in water. It is not destroyed by acids, alkalies, light, air, or heat. The word pellagra literally implies a rough red skin. The victim, known a pellagrin, has in addition to similar a dermatitis, a sore mouth. *Black-tongue* in dogs may be considered an approximate similar of human pellagra. While the dermatitis and accompanying inflammation of the tongue are the most outwardly characteristic symptoms of the disease, there are usually digestive and nervous or psychotic disturbances.

Elvehjem in 1937 showed that blacktongue in dogs can be prevented or (if not too advanced) cured by administration of nicotinic acid, much as beriberi is prevented and cured by thiamine. And within a few months at lest four medical research groups had exhibited the antipellagric value of nicotinic acid and of foods such as yeasts which are rich in it.

The inapprppriateness of the name nicotinic acid for a normal nutrient was recognized and *niacin* was coined as a name which in speaking or writing nutritionally may stand for the nutrient *factor* whether really present and functioning as nicotinic acid itself or its amide, or the chloride or hydrocholoride, or some conjugate.

Niacin : Its Functions

The primary role of niacin, as with thiamine and riboflavin, is as a constituent in coenzymes. Niacin is a constituent of two coenzymes that are essential for specific oxidation reactions to take place in the body. One of the chief functions of the coenzymes is the removal of hydrogen (dehydrogenation) from particular compounds and the transfer of it to others in a succession of complex reactions.

The Tryptophan-Niacin Relationship : It has been demonstrated that one of the essential amino acids, tryptophan, can be transformed into niacin in the animal body. This relationship between tryptophan and niacin was first reported in 1945 by Krehl and others at the University of Wisconsin, and since that time Reports have proved that human beings as well as most animals are able to modify the amino acid to the vitamin. Following this biological phenomenon, workers until 1945 had been unable to produce a human niacin deficiency by the deprivation of niacin alone in an otherwise required diet.

The transformation ratio of tryptophan to niacin in the human has been investigated. Horwitt and his group at Elgin State Hospital found the tranformation ration of tryptophan to niacin in adult men to be about 60 to 1. Using adult women as subjects, Goldsmith and coworkers at Tulane University have reported corresponding values to confirm the validity of this conversion ratio. Studies with pregnant and nonpregnant women point out that the conversion of tryptophan to niacin is more effective in the pregnant than in the nonpregnant state.

Niacin : Its Nutritional Chemistry

The normal nutritional functioning of niacin is largely, if not mainly, through the formation of compounds which act as coenzymes.

Like thiamine and riboflavin, niacin serves in the system of tissue enzymes inherent in the intermediary and oxidation-reduction metabolism. The complex interrelationships are still in process of being worked out. However niacin alone does not enable the typical pellagrin to become a healthy person, for the food supplies which have been accountable for the illness of most pellagrins are usually also deficient in one or more other nutrients. The metabolism of nicotinic acid as indicated by the urinary exertion of it and its derivatives has been studied by Perlzweig. It has been said that: "the major urinary excretory products in human subjects are 'N'-methylnicotinamide and N-methyl-6-pyridone-3 carboxylamide." There is cited a presumably typical experiment where in a subject on a "normal" diet excreted daily 6.3 mg. of N'-methylnicotinamide and 14.2 mg. of pyridone; after taking 100 mg. of niacinamide (by mouth), an excess of 26 mg. of N'-methylnicotinamide and 62 mg. of pyridone found in the urine;

while after taking 100 mg. of niacin (i.e., nicotinic acid) less of the end-products were recovered, 10 mg. of N'-methylnicotinamide and 26 mg. of pyridone. When 3 gm. of tryptophane were taken, 2.8 mg. of "extra" N'-methylnicotinamide and 16.6 mg. of "extra" pyridone were excreted— indicating a conversion of this amino acid into niacin or its products.

From similar studies it seems that, since of its more complex intermediary metabolism, niacin yields less clearcut consequences than do vitamin C and thiamine in studies of the concerning of nutritional intake, concentration level in the body, and urinary output. Nor does there seem to be eventual authority as to whether or not it is beneficial to maintain bodily saturation with niacin habitually.

Niacin requare not be supplied totally performed of particular precursor substances are available in enough quantity, for the human body which has tremendous ability to form niacin from precursors. Krehl and his coworkers have demonstrated the transformation of tryptophane to niacin in the rat. It has now been established that man, variou species of animals, microorganisms, and the growing corn plant, can fulfil this chemical change. The mechanism of this conversion, the site where it occurs-to what extent the body tissues accomplish it and to what extent it is due to the activity of intestinal microorganisms-still are not clear, but isotopic studies point out current progress. Methylation may take place and account for some of the recoverable end products.

As explained below, it is still not explicit to what extent the human requirement for niacin is met by the niacin formed by intestinal bacteria. It may be largely influenced by the diet.

Kruse (1942, 1942 b) studied the ascendancy of *aniacinosis* (niacin deficiency) upon the condition of the tissues, especially of the surface of the tongue. He devised a system of appreciating the condition of the tongue in aniacinosis which takes are of the form, intensity, and stage of the departure from normal. the two main categories of form are acute and chronic;

then the condition is further divided as to intensity or degree; and finally, as to the stage which the pathological process has reached. Kruse states that in practice it is possible to gauge intensity in three degrees; and stage in four steps if acute, or five if chronic. Also, that "if acute and chronic processes are both present, — as most frequently they are — they can be appraised separately" and then combined in a dual rating of status.

In niacin deficiency the tongue and its p^pillae show vascularity and hypertrophy or proliferation, followed by atrophy of the papillae. usually the fungiform precede•the filiform papillae in undergoing change. dissimilar sites on the tongue may show dissimilar stages, the usual order being: tip, anterior and antero-lateral edges, anterior and antero-lateral margins of the dorsum, anterior border of predominantly filiform zone, mid-dorsum. Late in the process, ulcers and erosions may occur on the tongue, usually beginning on the edges. Those siught a fuller account should read Kruse's original papers.

Niacin or its amide cured all these symptoms, the cures being quicker in the acute than in the chronic cases. When both acute and chronic pathology were present in the same person, the symptoms of the acute process disappeared first. Kruse streses the same principles here as in analogous cases of other avitaminoses, the chronic cases of long standing can be completely cured only by long-continued treatment. He thinks many patients may have been discharged with only their acute and not their chronic injuries cured.

The particular pathologic changes on the tongue are considered by Kruse (1942b) to be the earliest and most global and reliable symptoms of niacin deficiency and so of pellagra when hence defined. Kruse also speaks of this *glossitis* as the earliest recognizable tissue manifestation of niacin deficinency. He holds that, as the tongue symptoms usually appear previous than the dermaitis in the growth of pellagra, so similarly the dermatitis disappears previous than the glossitis in a cure. And he advices that too often the pellagra patient is discharged as cured when the skin symptoms are relieved whereas therapy ought to continue until the tongue symptoms also have completely disappeared which may requare a year or more.

Prominence of maize in the diet has been charged, particularly in European medical literature, with accountabilities for pellagra; but this is feasibly only true in the same sense that rice eating is concerned to beriberi, that is, too great dependence upon cheap ways of food calories which are but poor ways of the needed vitamin. most likely contributing to the influence also is the reality that maize proteins are notably low in tryptophane, which, as mentioned earlier, may serve as a precursor of niacin. That cornmela is not actively pellagra-producing when the rest of the diet is right, is reaffirmed by Kooser and Blankenhorn (1941) who found that one of two otherwise comparable communities studied has freed itself of pellagra by increased consumption of milk and eggs, and in lesser degree of lean pork and chickens, with little if any decrease in the consumption of cornmeal.

Other Factors Involved in the Pellagra Problem

Ball (1939), Elvehjem (1939), and S.G. Smith (1940) and some others have stressed, at least three vitamins of the B group (thiamine, riboflavin, and niacin) are collectively related in biological oxidations and any investigation of the role of one of these demands of consideration of the other two.

Independently, Sydenstricker, McLester, Spies, and others lay stress the common experience in pellagrous regions that while nicotinic acid cures the dermatitis and inflamed tongue which are the conditions to which the name pellagra partiularly applies, and usually also alleviates the digestive and nervous disturbances, yet the clinical pellagrin is not thereby rendered a completely healthy person. In fact, Sydenstricker (1941) construed pellagra as a combined deficiency of niacin and riboflavin. Multiple nutritional deficiencies seem, in the actual experience of American people, to be more frequent than deficiencies in niacin alone.

The unique diet of the poor pellagrin contains so little of anything else than milled cereals, sweets, and fats or fat meat that it is all too apt to induce other nutritional deficiencies at the same time with that of niacin.

Physicians dealing clinically with pellagra have repeatedly reported that while niacin cures the pellagra itself, several if not most clinical pellagrins requare also thiamine or riboflavin or both. And Spies has reported a group of cases who after receiving all three of these vitamins were still not entirely well, but showed further improvement after receiving pyridoxine (vitamin B_6) in addition.

What is most required in the pellagrous regions is a better food supply. Sebrell gives as the most important foods to add, "milk, liver, lean meats, fish, eggs, tomatoes, green peas, and a variety of green and leafy vegetables like kale, mustard greens, turnip greens, and collards."

The deficient diet of the pellagrous regions is, "due to by the cultivation of a money crop instead of food and forage crops." When even a small fraction of the time and land yet devoted to cotton is used for raising vegetables for the family and forage for a family cow, the health, and with it the earning power, of the people should be greatly improved.

The field studies of Sandels and Grady (1932) and of Stiebeling and Munsell (1932) both exhibited a vary clear connection between food habits and pellagra incidence.

It is gratifying to know that the educational work based on similar observations as the above, simultaneous opportunity for fuller employment and resultantly increased purchasing power have recently very greatly reduced the prevalence of pellagra. Fuller reporting of cases that still occur may tend to obscure the reality in the official statistics, but personal communications from Dr. V.P. Sydenstricker are conclusive that at least in a significant part of the South there has been a very real reduction in the prevalence of pellagra, attributable to improved food habits of the people in general and increased purchasing power of previously low-income families.

The Practical difficult of Pellagra Prevention

Largely privious the finding out of the concerning of niacin to pellagra, the United States Department of Agriculture had made comprehensive quantitative field studies of the relations of food consumption to pellagra incidence and the Public Health Service had experimented extensively with human subjects to find out the amounts of dissimilar foods that ample to prohibit pellagra. Bringing together the findings of these Government researches, it may be concluded that any one of the items of Table contains a full day's quota of pellagra preventive value:

Table : Food items Preventive of Pellagra

A quart of milk or buttermilk;
A pint of evaporated milk;
One third to one half pound
of dried skimmed milk,
of lean meat,
of canned salmon,
of peanut meal, or
of wheat germ;
One pound of fresh or canned collards,
of kale,
of green peas, or
of turnip greens; or
Two to three pounds of tomatoes, fresh or canned,
or of tomato juice.

Knowing the extreme carefulness of these governmental investigations, especially in the direct experimentation with human subjects. It is believed that the findings constitute important evidence. That for some foods it does not run parallel with the blacktongue preventive values derived from experiments with dogs may be partly becauese of pellagra as generally met being less purely a niacin deficiency than is blacktongue and partly to the reality that some foods are better used by human beings than by dogs.

With some foods there is, and with some there is not, a good quantitative close concerning between the more than direct authority of pellagra preventive value and the current analytical estimates of niacin content. This corresponds with other authority that present niacin assays, both "microbio" and *in vitro,* have more validity for some sorts of food than for others. These methods are fairly trustworthy for comparisons among meats, or among cereal products, or for control of enrichment in breadstuffs and cereals; but much less so for evaluations of widely divere sorts of food. Another complication lies in the reality that' foods may have both direct and indirect niacin values. In addition to the niacin that they really contain, some foods to a large extent— others to only a small extent, if at all—increase the body's niacin intake by favoring the activity of niacin forming bacteria in the digestive tract. If this is a significant factor in the niacin supply of human nutrition, it must also be subject to large individual variations, which would describe the reality that some patients do not derive the usual amount of pellagra-preventive value from certain foods.

The National Research Council's Recommended Allowances of niacin were set at ten times the imilar allowances of thiamine for both sexes and all ages. Obviously this was a tentative "first approximation." In 1951-52, however, it is still in force.

As yet, we do not know: (1) what part of the niacin required to be contained in the food as eaten, and to what extent a well-balanced diet will be "biologically enriched" in niacin by the bacteria of the digestive tract, or (2) the degree of trustworthiness of present estimates either of the niacin requirements of human nutrition or the niacin contents of some types of food.

The Impression of a Niacin Deficiency

In Animals : Only the dog develops characteristic deficiency symptoms since of a lack of niacin. Supposing that other animals display symptoms like reduced growth, loss of appetite, and weight loss, these conditions are only the standard symptoms observed in most vitamin deficiencies. Blacktongue, the term used to explain a niacin deficiency in a dog, manifests itself in serve lesions of the mouth, a characteristic halitosis, diarrhea, emaciation, and a lack of appetite for both food and water.

In Man : The first report regarding experimentally induced pellagra in man by dietary means was made by Goldberger and Wheeler in 1915. However, after the identification in 1937 of niacin as the antipellagra factor, innumerable workers endeavoured without success to produce experimental pellagra in man by niacin deprivation. It was not until 1952 Goldsmith and her associates at Tulane University were successful in inducing human pellagra by the utilize of corn diets low in niacin and tryptophan. Now it is known that these other workers who had failed to produce a human niacin deficiency had utilised experimental diets enough in all known nutrients except niacin, and that the stumbling block was the presence of sufficient tryptophan, which indirectly supplied niacin to the body.

The subjects of the study Gold smith developed pellagra during the fifty days they were ingesting a corn diet that was low in niacin and low in tryptophan. At the end of the study they were cured of the disorder with tryptophan, which further proved the importance of this amino acid in niacin metabolism. Prior to this work the intake of vitamins of the B-complex group other than niacin was recommended to improve the multiple lesions of pellagra. In this study, however, even though added amounts of these vitamins were present in the diet, the typical lesions of the disorder were produced. This observation indicates that pellagra is caused by a deficiency of one particular factor, niacin, and not by a group of factors.

In 1955 the same group of workers made a corresponding human experiment using a wheat diet instead of the typical corn diet that is associated with pellagra. It was found that the niacin deficiency took longer to develop and was less severe on the wheat diet than when the corn diet was utilised, even though the diets were comparable in niacin and tryptophan content. The authors indicate that the low tryptophan content of the corn may not be the only description for the pellagragenic influence of this cereal- It may be that the comparatively low digesibility of corn and the low availability of niacin in corn assist to the earlier growth of pellagra on corn diets.

Suggested Permission for Niacin

The Food and Nutrition Board (NRC) has expressed the allowances for niacin in concerning to of *niacin equivalents* (10). Equivalents of niacin comprise the dietary ways of the performed vitamin as well as the precursor, tryptophan; conversion ratio is 60 mg tryptophan equals 1 mg niacin.

The food intake on daily basis by a college or young adult man (18 to 35 years of age) should give 19 mg equivalents of niacin; a middle-aged man (35 to 55 years) should have 17 mg equivalents; and 15 mg equivalents have been proposed for older man (55 years and older). College and young adult women (18 to 35 years) require about 14 mg equivalents of niacin; those over 35 years have an allowance of 13 mg equivalents of the vitamin each day.

Niacin : Its Sources

Data on the pellagra-preventing potency of foods was reported by Goldberger and others long privious the recognition of niacin as a vitamin. This information was depended on studies of humans suffering rom pellagra and of blacktongue in dogs. Foods were rated as *good, fair,* and *little or none,* depending upon their curative influence of pellagra and blacktongue. Today, with the comprehensive information on the occurrence of niacin in foods, it has been displayed that this early rating system corresponds well to the niacin values measured by present-day procedures.

Although niacin iss found both in plant and animal products, brewer's yeast and peanut butter are the rihesst food ways of the vitamin. Glandular meats, muscle meat, poultry, and fish are rated as excellent niacin foods, whereas eggs are rated as niacin-poor. Milk, supposing that low in actual niacin content, must be grouped as a good antipelalagra food following its high tryptophan content. Fruits and vegetables vary in their content of niacin. Whole-grain cereals contain more of the vitamin than do highly refined cereals even though cereals are low in niacin. The enrichment of corn and wheat flours in this country has done much to increase the niacin content of our diets. A report from the U.S. Department of Agriculture has displayed that the national food supply in 1962 contributed about 28 per cent more niacin than it did for the period (1935-1939) prior to the adoption of cereal enrichment.

Coffee is a new item to be added to the list of niacin-contributors to the human diet. It has been reported that ordinary retail coffee contains about 10 mg niacin per 100 gm and instant powdered coffee about 30.6 mg per 100 gm. Explicitly the niacin level is based on the degree of roasting of the coffee beans: light roast equals about 1 mg niacin per cup, and dark roast equals 2.4 to 3.4 mg niacin per cup. In certain areas of the world where the diet is low in niacin the tryptophan, the high consumption of coffee may describe the low incidence of pellagra among these groups of people.

Niacin Equivalents of Some Typical Foods

At this time the niacin values in food tables are expressed in concerning the performed vitamin, milligrams of niacin, not in niacin equivalents. It is not yet realistic to express food values in equivalents of niacin due to the scarcity of sufficient data for the tryptophan content of various foods under dissimilar conditions of processing, storage, and preparation. In the 1963 edition of *Composition of Foods —Raw, Processed, Prepared* niacin values of foods are given regarding the performed vitamin.

To explain the contribution of the tryptophan precursor in foods, the niacin equivalents of some commonly used foods have been calculated (Table 9.2). The niacin assay of a cup of milk is listed as 0.1 mg. This same portion of milk consists of about 120 mg of tryptophan. Using the conversion factor, 60 mg tryptophan equals 1 mg niacin, a cup of milk contributes 2.0 mg of the vitamin from its tryptophan source. Thus a cup of milk contributes 2.1 mg equivalents of niacin to the diet, not 0.1 mg.

Niacin Losses During Food Processing and Preparation

Niacin, which is the most stable vitamin of the B-complex group, is not affected by the general factors encountered in food processing or storage. Following its water-soluble nature, however, a food may lose some of its niacin in the period of food processing or preparation. The utilize of a small amount of water in cooking niacin-containing foods will minimize the loss of the vitamin in the processss of cooking. The liquid fraction from the processing or preparation of a food should be utilise wish aview to reclaim any of the original niacin that was lost from the food by leaching.

Table : Approximate Niacin Equivalents of Some Typical Foods.

Food	*Measure*[1] *Serving*	*gm* (A)	*Tryptophan Content*[2] *mg/100 gm* (B)	*Niacin Equivalents from Tryptophan*[3] *mg/Serving* (C)	*Niacin Content*[1] *mg/Serving* (D)	*Niacin Equivalents* *mg/Serving* (C+D)
Beans, green, canned	1 c	239	14	0.6	0.7	1.3
Beef, ground	3 oz	85	187	2.7	5.1	7.8
Bread, white, enriched	1 sl	23	91	0.4	0.5	0.9
Carrot	1 med	50	10	0.1	0.3	0.4
Cheese, Cheddar	1 oz	28	341	1.6	trace	1.6
Chicken, broiler	3 oz	85	250	3.5	7.4	10.9
Corn, Canned	1 c	256	12	0.5	2.3	2.8
Egg	1 med	50	211	1.8	trace	1.8
Lettuce	2 lg lv	50	12	0.1	0.2	0.3
Milk, whole	1 c	244	49	2.0	0.1	2.1
Orange juice	1 c	249	3	0.1	0.8	0.9
Peanut butter	1 tbsp	16	330	0.9	2.4	3.3
Peas, green, canned	1 c	249	28	1.2	2.2	3.4
Potatoes, white	1 med	136	18	0.4	2.0	2.4
Salmon, canned	3 oz	85	200	2.8	6.8	9.6
Tomato	1 med	150	9	0.2	1.0	1.2
Yeast, brewer's	1 tbsp	8	710	0.9	3.0	3.9

The Enrichment of Cereal Grains

About 72 per cent of the Calories consumed by the people of the world come from rice, wheat, corn, and other grains. Wheat and rice, however, due to consumer preference and a better keeping quality, are milled to a highly refined form causing a loss of the nutrients that are vota; in regulating the body's utilize of this food energy. Apart from diets rich in corn products may be pellagra-producing due to the inadequate amounts of niacin and tryptophan in this cereal food. In order to combat the clinical and subclinical deficiency diseases which may consequence from insufficient amounts of the B-complex vitamins in cereal-rich diets, the enrichment of wheat, corn, and rice products.

The committee on Food and Nutrition of the National Research Council, now called as the Food and Nutrition Board, deserves much credit for the adoption of cereal enrichment in this country. Supposing that the enrichment of white flour and bread began in 1941 under a War Food Order of the Federal Government, South Carolina in 1942 became the first state to enact legislation which demanded enrichment of all wheat products sold within its boundaries. After the final opinion of the Second World War, when wartime directives ceased to exist, the idea of enrichment was well established as nineteen states had passed legislation in connection wish it. In 1962 thirty states and Puerto Rico had passed enrichment laws, and in the other states about 92 per cent of the white bread sold was enriched. In five southern states, where cornmeal and grits are used in large amounts, legislation has been extended to comprise the enrichment of these products. Only one state, South Carolina, has enacted legislation to include the enrichment of rice.

Thiamine, riboflavin, niacin, and iron are the four nutrients that must be added to white flour or bread under the enrichment law. The addition of vitamin D and calcium are optional but not mandatory. In order for one to see the value of the enrichment of wheat flours, a comparison of the vitamin and iron content of all wheat, enriched white, and unenrichment white flours in displayed. Enriched white flour conssists of about twice as much riboflavin but only four-fifths as much thiamine, niacin, and iron as is found in all wheat flour. On the other hand, enriched flour consists of nine times as much thiamine, six times as much riboflavin, and about four times as much niacin iron as does unenriched flour. The addition of these nutrients to white flour reasons no modify in taste, color, or cooking quality. The present cost to enrich wheat flour is about 3.6 cents per hundred pounds.

The value of white flour enrichment to the nutritional well-being of a population has been displayed in the popular Newfoundland surveys (1,2). privious the introduction of white enriched flour into Newfoundland in 1944, medical surveys were made in two areas of the country. Four years subsequent a resurvey was made in the same areas to condition the profits derived from similar a program. The clinical signs of malnutrition, following the period of the addition of thiamine, riboflavin, and niacin to the diet through the enriched flour, had decreased. By 1948 the common health of the people had improved, but there was no improvement in the lesions related to ascorbic acid malnutrition where no dietary supplement of that nutrient had been provided.

The diet of the people of the United States were improved significantly by enrichment of bread and flour. Lower income families benefited relatively above higher income families since of a greater consumption of flour and bread.

Rice can be enriched by coating it with a mixture of thiamine, niacin and iron, and this mixture is called *Premix*. When Premix is utilised in amount of 1 part to 200 parts of rice, the enriched product has a nutritive value comparable to brown rice. In the past riboflavin was not used in rice fortification since it discoloured the grains. Now a new process has been developed that overcomes this yellow spotting of the rice grains.

The peoples of Asia, who forms all but one-half of the world's population, eat about 80 to 90 per cent of their Calories in the shape of rice. The nutritional scarcity of rice as a staple food are explicit in the high incidence of beriberi among the peoples of this continent. The solution for the control of beriberi in the Far East appears to be in the enrichment of rice — demonstrated by studies conducted in the Bataan Province of the Philippine Islands. Nine months after the introduction oı enriched rice to the seven experimental towns it was found that nearly 90 per cent of the people who had previous displayed mild to definite beriberi symptoms were significantly improved. The incidence of death due to beriberi was reduced by 67 per cent during the first year of rice enrichment, and not one death since of this disease was reported during the latter quarter of the second year. Through the endeavours of Dr. R. R. Williams of the Williams-Waterman Fund (8,9) and other scientific and civic leaders, both in the United States and in the Philippines, a plan was started in 1955 that brought the entire Philippines under rice enrichment by law late in 1956.

8

Pyridoxin : Its Nutrition

More over thiamine, riboflavin and niacin there are other factors of the B-complex family that qualify as vitamins. Due to this cause investigators have used both deficiency diets and chemical materials known as antagonists. As most of these factors act as coenzymes in the enzyme systems of the body, they may be approach of as quite similar in their function. Because these vitamins are required in minute amounts and are found in the commonly used plants and animal foods of the diet it is not likely that human beings will ever have a deficiency. usually the ordinary practices of food processing and food preparation do not profoundly assume the content of any of these vitamins in the food intake. Vitamin B_6, folacin, pantothenic acid, and vitamin B_{12} are necessary in human nutrition and must be involved in the daily diet. A human dietary requare of biotin, choline, and inositol is doubtful at this time. Choline can be synthesized in the body from many substances, and biotin is synthesized by bacteria tract.

Vitamin B_6

Another factor in the B_2 complex was recognized in 1934

as being distinct from riboflavin. This was B_6. It may be wondered why it was not B_3, B_4 or B_5 but, as stated before, these symbols had already been used for other substances approach at the time to be new vitamins. The terms were dropped when it was proved that they were not separate entities.

B_6 deficiency caused a sort of dermatitis in rats (the so called rat pellagra). A number of chemically concerned substances have B_6 activity but the commonest form is pyridoxin, present with vitamin B_1 in relatively high concentration in wheat germ. There has never been a unit of B_6 activity since by the time that sufficient was known about it to requare measurement in foods, the pure materials was available for comparison.

Vitamin B_6 was the first designation of a then unidentified factor needed for development and for prevention of dermatitis in the rat. As the dermatitis caused by its lack is, in the case of the rat, particularly striking, vitamin B_6 was sometimes also called "the antidermatitis vitamin" and confused with the pellagra-preventive factor. Soon after the latter was detected as nicotinic acid, vitamin B_6 was provided the name *pyridoxine* (adopted during the spring of 1940 by the American Institute of Nutrition, the American Society of Biological Chemists, and the Council of Pharmacy and Chemistry of the American Medical Association).

Spies, as we saw, reproted indications of a vitamin B_6 deficiency disease generaling as a complication of pellagra; but Elvehjem has designated it as a questionable factor in human nutrition. Its role in nutrition and its therapeutic probability are difficulties for further research.

The apparent relationship of pyridoxine as observed in laboratory experimentation is to the prevention of the acrodynia-like dermatitis in rats. There is some authority that both vitamin B_6 and a fatty acid or fat-soluble factor are concerned to the prevention of this dermatitis; and the latter factor shows resemblances to the necessary fatty acid of Burr and Burr and to the fat-soluble antidermatitis factor of Hogan and Richardson.

In another aspect of its functioning, vitamin B_6 occurs in many body enzyme systems included in the metabolism of proteins and amino acids.

Varied recent studies have advised that pyridoxine is required by man and that blood levels and urinary exertion development and fall with intake. A naturally occurring deficiency has not been explicitly displayed but the therapeutic usefulness of pyridoxine in the nausea of early pregnancy has been concerned to a possible insufficiency of this factor in similar cases.

Pyridoxamine and *pyridoxal* are distinct but chemically concerned substances which also have vitamin B_6 activity. There is now a tendency to prefer the term "vitamin B_6" where the group of active substances is met, and reserve chemical designation for a single certain substance. This factor was named pyridoxine by Gyorgy in 1934. However, there are three correlated chemical compounds with potential vitamin B_6 activity — pyridoxine, pyridoxal, and pyridoxamine (Fig. 1). The pyridoxal and pyridoxamine forms occur primariliy in animal products whereas pyridoxine is found on comprelensive bassis in products of vegetable origin.

Vitamin B_6 : Its Functions

Vitamin B_6 functions as part of coenzyme systems in the body that assist in the metabolism of amino acids, fatty acids, and in the release of energy. It is found predominantly in tissues as pyridoxal or pyridoxamine phosphate.

Pyridoxal phsophate (vitamin B_6 coenzyme) functions in so several reactions that are necessary in the metabolism of amino acids: decarboxylation (removal of CO_2 group), transamination (transfer of NH_2 group), and desulfuration (removal of H_2S group). The transformation of the amino acid, tryptophan, to the vitamin, niacin, bases on the action of this coenzyme. There seems to be a relationship between vitamin B_6 and necessary fatty acid metabolism. Witten and Holman postulate that vitamin B_6 is liable for the transformation of linoleic acid to arachidonic.

A pyridoxine deficiency has been found to assume antibody formation in man and animals. Hodges and associates reported that an induced pyridoxine or pantothenic acid deficiency resulted in impairment of the human antigenic response to tetanus and typhoid immunizations.

The Impression of a Vitamin B_6 Deficiency

In Animals. All types of species of animals appear to require vitamin B_6 for normal growth, yet dissimilar kinds of symptoms consequence when different species of animals are deprived of it. Supposing that the rat loses weight very rapidly on a vitamin B_6-deficient ration, the characteristic lesion in this species is a skin disorder known as *acrodynia*, which may develop on the tail, paw,nose, mouth, or ears (Fig. 2). As the deficiency advances, however, the rat may suffer from muscular weakness and convulsions.

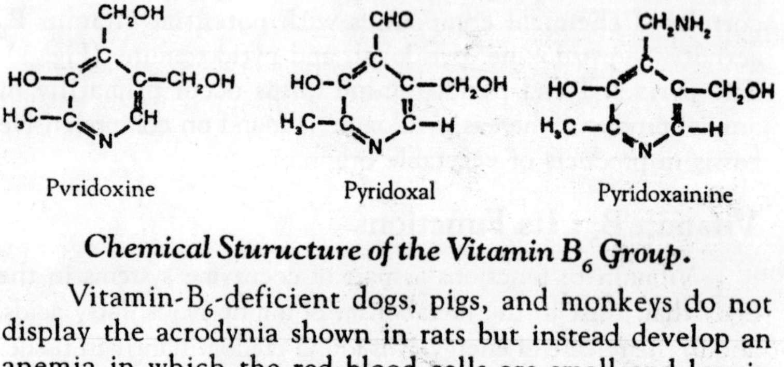

Pvridoxine Pyridoxal Pyridoxainine

Chemical Sturucture of the Vitamin B_6 Group.

Vitamin-B_6-deficient dogs, pigs, and monkeys do not display the acrodynia shown in rats but instead develop an anemia in which the red blood cells are small and low in hemoglobin (microcytic hypochromic anemia). Notwith standing in the subsequent stages of the deficiency, these animals too may suffer from convulsions. Neither the acrodynia nor the anemia is a characteristic symptom of calves deprived of vitamin B_6. These deficient animals show anorexia, listlessness, a poor coat of hair, and sometimes convulsions.

In man. The human requirement for vitamin B_6 has been established from deprivation studies of both adults and infants. It was necessary to resort to experimentally induced vitamin B_6 deficiency to establish the requare for the vitamin. So widespread is the occurrence of vitamin B_6 in foods that a deficiency does not comes out naturally. The three symptoms found in these human experiments where skin lesions, anemia, and convulsive seizures.

To study the influence of vitamin B_6 deprivation in adults, eight subjects were fed a low vitamin B_6 diet as well as desoxy-pyridoxine, the antagonist of the vitamin. After two to three weeks on this diet seven of the subjects developed a greasy scaliness on the skin about the nose, mouth, and eyes (Fig. 3). These lesions remained unchanged when thiamine, riboflavin, and niacin were ingested but disappeared in several days after the administration of vitamin B_1.

The influence of a vitamin B_6 deprivation is more dramatic in the infant than in the adult. The two infants that were studied experimentally by Snyderman and her group stopped receiving weight some weeks after being fed a diet deficient in the vitamin. Supposing that one of the infants developed convulsive seizures after 76 days on the diet, the other one made no convulsive activity pattern but did develop anemia after about 130 days of vitamin deprivation. Both of these conditions, however, were rapidly relieved by the administration of pyridoxine. Between 1951 and 1953, reports came from dissimilar parts of the United States describing symptoms of a nervous irritability and convulsive seizures in infants less than 6 months old. It was found that all the babies affected had been fed the same commercial, canned, liquid-milk formula. A history of the symptoms and treatments of this disorder in a 2½ months went into convulsive seizures. Three minutes after the baby had received an intramuscular injection of 100 mg of pyridoxine hydrochloride, the convulsions had stopped and the child's pattern of activity was normal. It is now known that the reason of these convulsive seizures was a short age of vitamin B_6 in the diets of the infants fed that particular preparation since the natural vitamin B_6 of the milk formula had been destroyed by the heat processing in the process of its manufacture.

The only utilize of non-fat milk (skim) powder in infant diets might also produce a vitamin B_6 deficiency since of the destruction of the vitamin during processing. Because of th experience, processed milk formulas are now analyzed and those on the market consit of a enough amount of the vitamin. A daily allowance of 400 ug of vitamin B_6 has been advised for the artificially fed infant.Although vitamin B_6 relieves the convulsive seizures resulting from a deficiency of the nuttrient, it has been exhibited to be of no value in the control of convulsive seizures in the treatment of epilepsy in children.

There are other abnormalities that may be concerned to a deficiency of vitamin B_6 in animals. Several studies endorse the finding that a deficiency of the vitamin favours the deposition of kidney stones (calcium oxalate) in animals and man. specific sorts of human anemias have been observed that respond to pyridoxine therapy that the vitamin may offer some protection against dental caries during human pregnancy has been suggested also by the study of Hillman and his associates.

Vitamin B_6 : Its Sources

Vitamin B_6 is amply distributed among the plant and animal foods of the diet. Supposing that the best ways of the nutrient are found in muscle meats, liver, vegetables, and whole-grain cereals, few foods can be divided as poor ways of vitamin B_6. For example a slice of whole-wheat bread (0.096 mg) consists of about four times as much of the vitamin as does a slice of white bread (0.023 mg). Milk, a widely used food in the American diet, contributes about 0.175 mg of vitamin B_6 per pint.

Suggested Permission for Vitamin B_6

The human allowance for this vitamin has not been definitely established but at this time a daily intake of 1.5 to 2.0 mg per day has been advised as an approriate allowance for the adult. As vitamin B_6 is widespread in both plant and animal foods, this amount will be given in the average mixed diet. It has been reported that in high and low-cost enough diets and even in poor diets in the United States, the vitamin B_6 content about 2 mg., 2.7 mg, and 1 mg, respectively, per day.

9

Folic Acid : Its Nutrition

Pteroylglutamic acid is known as Folic acid. *It is also known a* *Vitamin* M. Folic acid, as designated for the first time as vitamir M, is now chemically accepted as pteroylglutamic acid (o glutamates); vitamin B_{12} is now held to be the specificall' antipernicious anemia substance; and the citrovourm factoı into which folic acid is changed *in vivo* in the animal organism is a materials needed for development by the bacteriun *Leuconostoc citrovorum.*

Folacin (Folic Acid)

Long privious folacin was either isolated or synthesizec its deficiency symptoms had been mentioned in man, animal: and microorganisms. Many laboratories had independentl studied the influence of a folacin deficiency in a number c species, and dissimilar names for this nutrient such as vitami M, vitamin B and L. *casei* factor had been reported. Now it known that there is a succession of biologically activ derivatives of folacin (pteroylglutamic acid) that make up th folic acid group of vitamins. Members of this group includ folacin, pteroyltriglutamic acid, pteroylheptaglutamic acid, an folinic acid or citrovorum factor, a derivative of folic acid (9). It appears that folinic acid is the biologically active form of the vitamin.

For the first time it was found out in 1941 (Mitchell and others, 1941) as a factor in green leaves (folium = leaf) needed to protect chicks from a different sorts of anaemia. Later similar factors were evaluated which were needed for the growth" of microorganisms. These were grouped together as folic acid when an active compound was synthesized in 1945.

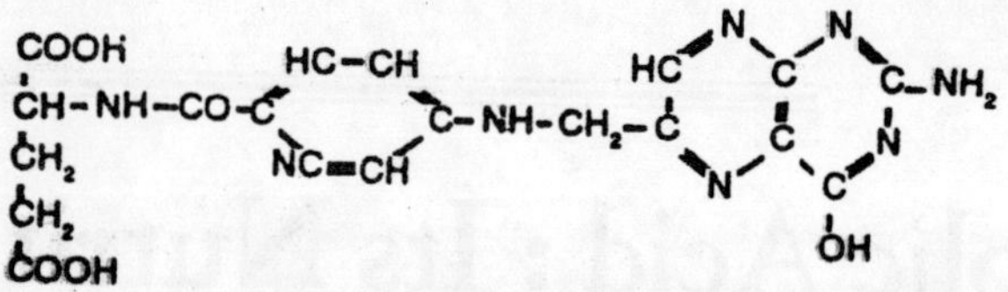

Previously unknown materials discovered by Day, Langston, dar by and lowoners to be required in the nutrition of the monkey. This they designated *vitamin* M. Other groups of workers used folic acid, vitamin B_c, and still other designations for what finally proved to be necessarily the same materials in antianemic activity. Chemically it was found to be pteroylglutamic acid and its formula was established as

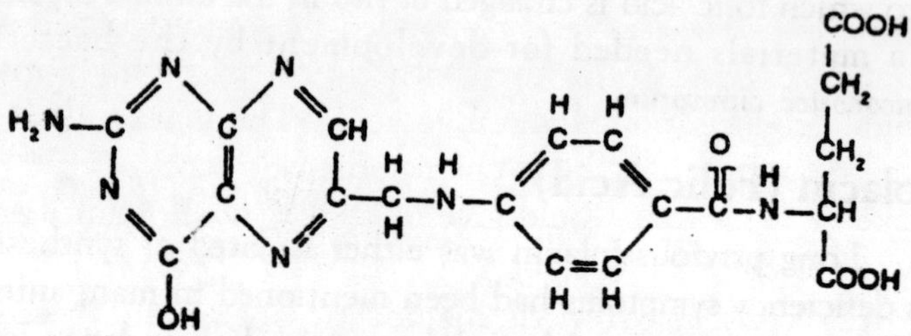

The B vitamins are of interest in having another feature in common. Several of them can be synthesized by bacteria and other micro-organisms like moulds. It is normal for particular kind of bacteria to inhabit the intestinal tract of man and other animals. These reason no harm and must not be ·fused with disease producing bacteria or germs. Some of these receive their own nourishment from the intestinal contents of their host and in turn they synthesize B vitamins, which hence become available for the host. this process reaches its highest growth in ruminants like the cow. The rumen comprise's a large volume of semi-fluid material at a temperature ideally proper for the bacteria to function and significant quantities of some of the B vitamins are manufactured there.

Vitamin B_1, riboflavin, nicotinic acid and pyridoxin are all synthesized in this source and also in other sites in the intestinal tract. Vitamin B_{12} seems to be the only vitamin known to be synthesized only by micro-organisms. It is not found, even in little quantities, in plants.

Various Steps of Folic Acid's Nutrition

In 1948 King reviewed in Reprint 129 of the National Research Council Reprint and Circular succession three lines of authority that this materials (folic acid, pteroylglutamic acid, vitamin B_c, L. *casei* factor, or vitamin M) is necessary to normal human nutrition.

(1) Innumerable widely distributed instances have been found in which individuals who revealed no other basis for functional impairment characterized by macrocytic anemia and concerned tissue changes, responded promptly to folic acid therapy, applying either the synthetic or the highly purified natural product (pteroylglutamic acid). The dietary histories of these patients have been characterized, in most cases, by an intake of foods that would give only very small quantities of folic acid.

(2) The characteristic macrocytic anemias accepted in infants, children, adult women, and adult men, most logically interpreted regarding an etiology of folic acid deficiency, have been strikingly similar to conditions established by folic acid deficiencies in controlled animal experiments.

(3) Efforts to deplete volunteer subjects of their tissue reserves of folic acid by supplying diets sufficient in other necessary nutrients but low in folic acid have met with dissimilar

success. Some investigators have not found a characteristic depletion in blood, urinary, or fecal concentrations of folic acid, but in other examples there has been authority of a slow depletion, like would be anticipated on the ground of animal experiments and clinical observations of individuals in an uncontrolled environment. Synthesis of folic acid by micro organisms in the intestinal tract apparently accounts for a major part of the irregularity in depletion experiments. The incidence of clinical deficiencies, however, affords evidence that intestinal synthesis should not be depended upon as a way of this nutrient.

"The quantitative requare cannot be estimated closely from published evidence. Doan, et al. observed a hematologic response from a dosage of 2 mg. of synthetic folic acid per day; Kurnick observed improvement after a dosage of 2.5 mg. per day of folic acid; and Moore, et. al. observed a response from 2 mg. per day.... Comparative studies with monkeys, chickens, and turkeys, on a caloric basis, given an indirect basis for estimating a human requirement in the range of 0.1 to 0.2 mg. per day. A number of investigators suggest that estimates of requare should be based upon studies of nutritional macrocytic anemias and directly concerned cases readily than upon protection from pernicious anemia."

Folic acid is active hematologically in macrocytic anemias, but it pernicious anemia neither prevents nor alleviates neurological symptoms. However, in macrocytic anemias of infancy and prequancy nutritional macrocytic anemia, and in sprue it is a highly effective therapeutic agent.

Distribution of folic acid

R.J. Williams and coworkers say that the available data on quantitative distribution of folic acid " are too low by a variable and unknown amount" since when these data were conditioned enzymes capable of freeing folic acid from tissues were not known.

In recent past however, Toepfer (U.S. Department of Agriculture), in work more particularly directed to the quantitative aspect, has conditioned the folic acid content of many foods, representative cases of which are involved.

Table : Total Folic Acid Content of Foods

Approximate averages or representative values: milligrams per kilogram (parts per million)	
Almonds	0.45
Apples	0.01
Asparagus (Fresh)	0.86
Bananas	0.10
Beans, green	0.23
Lima (fresh)	0.34
Wax	0.36
Beef, round, I	0.07
Beef, round, II	0.09
Beef, round, III	0.10
Beets	0.16
Bran	1.0
Brazil nuts	0.05
Broccoli	0.33
Cantaloupe	0.04
Carrots	0.06
Cauliflower	0.13
Celery	0.08
Cheese (Cheddar type)	0.12
Chicken, light meat	0.03
Chicken, dark meat	0.03
Chuck(Texas study)	0.15
Corn, sweet	0.14
Eggs	0.03
Filberts	0.71
Heart	0.03

Kale	0.42
Kidney	0.58
Lamb	0.03
Liver	2.94
Milk (butter milk)	0.11
Mustard greens	0.38
Oatmeal	0.31
Oranges	0.05
Peanuts	0.61
Peas, green, fresh	0.25
Peas, dry blackeye	4.9
Pecans	0.34
Potatoes	0.06
Pumpkin	0.04
Tomatoes	0.04
Walnuts	0.78
Wheat, entire, I	0.45
Wheat, entire, II	0.41
Yams	0.19

Table : Vitamin B_6, Pantothenic Acid, Biotin, Folic Acid, and Choline Content of Selected Foods[1]

Food	*Protion Amt*	*Wt, gm*	*Vit B_6, mg.*	*Pantothenic Acid, mg.*	*Biotin, mg*	*Acid, mg.*	*Folic Choline, mg*
Bread, white	1 sl	23	23.0	0.101	0.3	3.4	—
Bread, all wheat	1 sl	23	96.0	0.182	0.4	6.9	—
Cheese, Cheddar	1 oz	30	18.7	0.114	1.0	4.5	13.6
Egg	1	50	126.0	0.795	11.2	2.5	252.0
Milk, whole	1 c	244	87.8	0.756	11.5	1.5	36.6
Apple	1	130	39.0	0.130	1.2	2.6	—
Banana	1	100	320.0	0.310	4.4	9.7	—
Orange	1	100	31.0	0.220	1.9	5.1	—
Apricots, canned	½ c	125	67.5	0.125	—	0.6	—

Pineapple, canned	Ke	125	88.7	—	—	1.0	—
Beef, liver	2 oz	57	378.5	5.324	54.7	167.6	290.7
Beef, round	3 oz	100	495.0	0.520	2.6	10.5	68.0
Ham	3½ oz	100	440.0	0.640	5.0	10.6	122.0
Chicken, dark		100	25.0	0.692	10.0	2.8	—
Chicken; white	3½ oz	100	130.0	0.804	11.3	3.0	—
Tuna, canned	3½ oz	100	670.0	0.420	3.0	1.8	—
Beans, Lima	3½ oz	100	170.0	0.450	—	34.0	—
Beans, snap, green	3½ oz	100	63.0	0.200	—	27.5	42.0
Broccoli	2/3 c	100	171.0	1.290	—	53.5	-
Carrots	½ c	55	66.0	0.150	1.4	4.4	7.0
Peas, green	3½ oz	100	150.0	0.820	9.4	25.0	75.0
Spinach	3½ oz	100	198.0	0.310	6.9	75.0	22.0

Folacin : Its Function

The folic acid group works as coenzymes in the transfer of one carbon unit which is potent in the metabolism of many body compounds. Purines and pyrimidines, instances of these sorts of compounds, are utilized in the building of nucleoproteins and are found in the nucleus of every cell. Folacin is must for normal blood formation (hematopoiesis) due to its presumed function in the formation of purines and pyrimidines. In a study of megaloblastic anemias Vilter reported that "all authority points toward a dominant chemical role for the folic acid coenzymes in hematopoiesis, with vitamin B_{12} and ascorbic acid playing dependent parts."

Table : Average Content of Selected Nutrients in Various Categories as Calculated from Food Composition Tables and Determined by Microbiologic Assays

Nutrients per Day	*High Cost Calculated*	*Low Cost Analyzed*	*Poor Calculated*	*Analyzed*	*Calculated*	*Analyzed*
Calories	3013		2439		1117	
Protein, gm	126	130	112	116	32	31
Folic acid, ug	120.4	193.4	117.0	157.4	30.8	47.3
B_6, mg	—	2.0	—	2.7	—	1.0

Pantothenic acid, mg	11.0	16.3	8.9	14.2	4.1	6.0
B_{12}, ug	—	31.6	—	16.0	—	2.7

Adapted from Mangay Chung et al. Am. J. Clin. Nutr. 9:578, 1961.

The Impression of a Folacin Deficiency

In Birds and Animals. Supposing that all animals have a requirement for folacin, the amount required in the daily ration vanging from species to species, as some of them can produce the vitamin by intestinal synthesis. The dog, rabbit, and rat meet their requare for the vitamin by intestinal synthesis, but chicks and monkeys must have folacin in their feed, otherwise a deficiency will occur.

In Man : Folacin is effective in the relief of particular types of human macrocytic anemias. in order to show better the reason and impression of this sort of anemia, a brief description is provided of the growth and development of the red blood cells. After the red blood cells are yielded in the marrow of the bone, they must undergo a process of development and growth called *maturation* privious the cells are discharged into the blood stream ready to carry on their function in the body. In the first stages of their development, the immature red blood cells are large, nucleated, and include little or no hemoglobin. A *megaloblast* is an instance of this sort of immature cell. As maturation proceeds, the nucleus becomes smaller, more and more hemoglobin is formed, and the size of the cell is reduced. The mature red blood cell is called an *erythrocyte.* The process of maturation is dependent on the action of specific enzyme systems, and it is believed that it is in this role, as part of a coenzyme, that folacin functions in hematopoiesis.

A person suffering from a macrocytic anemia caused by a folacin deficiency shows a characteristic blood pattern as well as other typical symptoms of the disorder. In the blood stream there are many large, immature megaloblasts and relatively few mature erythrocytes. Following this condition, the oxygen-carrying capacity of the cells, which have not formed substantial amounts of hemoglobin, is greatly reduced and consequences in symptoms of weakness, quick breathing, and a slowing down of the body processes. The normal development and growth of other blood factors, the white blood cells *(leucocytes)* and blood platelets *(thrombocytes)* are also changed during a folacin deficiency. A decrease in the leucocytes, which are the body's chief defense against micro-organisms, will consequence in less resistance to disease on the part of the aneima patient, and his normal process of blood coagulation may be affected by the lowered number of thrombocytes. Apart from the impression caused by an alteration of specific blood constituents, the macrocytic anemia patient may exhibit weight loss, inflammation of the tongue, and disturbances of the intestinal tract.

Folacin has been displayed to be effective in relieving the symptoms in patients with tropical and nontropical sprue, nutritional macrocytic anemia, macrocytic anemia of pellagra, megaloblastic anemia of pregnancy, and megaloblastic anemia of infancy. Also this nutrient has been shown to improve hematopoiesis in pernicious anemia but "it is an incomplete treatment for this disease." Folacin has no influence whatsoever in relieving the neurological manifestations characteristic of pernicious anemia; actually it seems to aggravate the neurological lesions that involve the degeneration of the nerve cord. The macrocytic kind of anemia does not respond to folacin therapy.

That combined deficiencies of folacin, vitamin B_{12}, and ascorbic acid are much more general than a single deficiency of any one of these nutrients has been proposed. Vilter et al. believe that "the general defect liable for a megaloblast is an abnormality in folic acid metabolism which reduces the amount or the activity of the folic acid coenzymes. Deficiencies of vitamin B_{12} and ascorbic acid adversely assume formation of these coenzymes..."

Varied types of antagonists of folacin has been recognised. When these antivitamins, like aminopterin, are fed to animals, one of the early alterations in the process of hematopoiesis is a reduction in the number of leucocytes or white blood cells produced. Due to this finding, the administration of folacin antagonists has brought temporary relief, but not a cure, in cases of human leukemia, a disease in which there is a marked increase in the production of leucocytes.

Folacin : Its Sources

Folacin is extensively distributed in nature, particularyly in the green foliage of plants; thus the name folacin. The richest food ways are found in chicken livers and vegetables like asparagus, broccoli, endive, leaf lettuce, and spinach. Liver, legumes, and other leafy-green vegetables are good supporters of the vitamin.

Suggested Permission for Folacin

So far the National Research Council has not advised a folacin allowance for man. However, based on the folacin analysis of substantial diets (0.193 and 0.157 mg per day) and insufficient diets (0.047 mg per day) in the United States, it has been stated that a ".... diet comprising 0.15 mg total folic acid activity should supply at least 0.05 mg folic acid which is active for human beings."

Since folic acid in amounts greater than 0.1 mg per day mask the neurological symptoms of pernicious anemia, the sale of the vitamin in absence of prescription in amounts recommending doses greater than 0.1 mg per day is precented in the United States.

10

Vitamin B_{12}: Its Nutrition

Minot and Murphy in 1926 found that pernicious anemia responded to treatment with all liver or its extract. Until 1948 that scientists isolate could not be able to a red crystalline material that in very small amounts was advantageous in the relief of this form of anemia. The substance, which is the most recently invented vitamin, was named vitamin B_{12}. It is a red crystalline compound with a complex formula ($C_{63}H_{88}N_{14}O_{14}PCo$) containing two minerals, cobalt and phosphorus. Vitamin B_{12} is not a single materials but may exist in several dissimilar forms. Vitamin B_{12} (cyanocobalamin) and vitamin B_{12a} (hydroxocobalamin) are the two active members of this group.

It is the new inventation in this group, and is one of the most highly active substances known. The first vitamin B_{12} to be obtained in the pure state was extracted from liver. It is now prepared on a large scale by different process. Its structural formula has not yet (1954) been completely elucidated. A readily unusual feature is that it contains cobalt. It is deep red in colour.

Historical Backdrop

Minot and Murphy in 1926 found that pernicious anemia responded to treatment with all liver or its extract, and during 1927 to 1980 West and Dakin worked actively on the fractionation of liver extracts, testing the fractions in actual therapeutic utilize with pernicious anemia patients whenever similar patients were available. Other research teams also had worked actively in this same field for several years when, in 1948, Smith and Parker in England, and Rickers, Brink, Koniuszy, Wood, and Folkers in the United states were to gether successful in the isolation of a crystalline red materials comprising cobalt and phosphorus, which was completely effective (in correct conditions) against pernicious anemia. The pure substance, designated vitamin B_{12} in 1948, was found to be so active therapeutically that a single dose of 3 to 6 micrograms was clinically effective when injected intramuscularly in patients with pernicious anemia. Clinical tests have exhibited that vitamin B_{12} is completely effective in the treatment of pernicious anemia, arresting central nervous system degeneration as well as restoring normal blood values.

Vitamin B_{12} : Its Various factors

The complete effectiveness of vitamin B_{12} in pernicious anemia led to intensive study of its place in the postulated theory: extrinsic factor + intrinsic factor—erythrocyte maturation factor; and the role of extrinsic factor has been assigned to vitamin B_{12}.

Berk, et al. (1948) found that vitamin B_{12} administered verbally in amounts effective when injected was hot effective unless accompanied by a way of intrinsic factor, e.g., normal gastric juice, but that if very large amounts were fed, it was effective alone. This advised the recognition of vitamin B_{12} with the extrinsic factor, a theory which has been sustantiated by the work of diverse groups approaching the difficult from dissimilar angles. Other ways of extrinsic factor, such as beef muscle, have been shown to be effective in proportion to their vitamin B_{12} content.

The role and recognition of intrinsic factor is still, however, an unsolved problem. It has been suggested that the function of intrinsic factor is to facilitate the absorption of vitamin B_{12} in some manner. The finding of a materials in normal gastric juice that combines with vitamin B_{12}, formulating it unavailable for the development of micro-organisms, has been reported by various. Ternberg and Eakin have isolated from gastric juice and from hog mucosa a non-dialyzable, heat-labile material, apoerythein, which combines stoichoiometrically with vitamin B_{12}. The resulting complex, erythein, is inactive for micro-organisms, releases vitamin B_{12} on heating, and is highly selective in its binding action. Vitamin B_{12} is bound by other substances also, live the lysozyme of eggwhite, but there is a distinction in the complex formed with apoerythein in that vitamin B_{12} cannot be separated from it by dialysis, a ready means of separating it from other protein substances.

A study of the reactions between normal gastric juice and vitamin B_{12} by Wolf, et al. indicates that supposing that it enters into sort form of combination, there is no great modify in the vitamin itself, and hence strengthens the theory that the function of the intrinsic factor is to foster absorption of vitamin B_{12}, the extrinsic factor.

Vitamin B_{12} and Growth

Vitamin B_{12} has been found to have growth-promoting activity for some micro-organisms, for animals, and in some cases for children.

Wetzel and coworkers (1949) have presented strikingly favorable consequences from vitamin B_{12} in 5 of group of 11 children who had been formulating less than normal growth. Supposing that there were no characteristic or suggestive physical signs in the hair, skin, eyes, mouth, or nervous system prior to offering vitamin B_{12} the growth response was supported by improved vigor and alertness, demonstrating that this vitamin has a far more common nutritional significance than its anti-anemic influence alone.

The development promoting property of the so-called "animal protein factor" has been reported to be due in part to vitamin B_{12}, but other impression of this factor, which appears to be of a multiple nature, are due to substances which await the consequences of further research for identification.

Conspicuously vitamin B_{12} occurs in at least two active reversible forms that vary in their influence on the development of micro-organisms, supposing that they are both available in analytical feeding tests. The vitamin is known to be formed by several of the lower plants, comprising specific bacteria, yeasts, and molds. In view of this, it now becomes quite gratuitously confusing to link it with the misnomer of "animal protein factor." Rather may this latter term be dropped in the interest of clarity, both here and in comparisons of proteins.

Vitamin B_{12} folic acid, and ascorbic acid

Vitamin B_{12} and folic acid have been shown to have their hematological impression increased in specific cases of stuborn megaloblastic anemia by the administration of ascorbic acid. In a deficiency of ascorbic acid in infants, larger amounts of folic acid are needed for the remission of anemia. A dearth of response to vitamin B_{12} in megaloblastic anemias has been observed by many workers, success being achieved with folic acid alone. It was reported by Holly the successful treatment of similar anemias by the combined utilize of ascorbic acid and vitamin B_{12}. Studies of Sealock, Day, Darby, and their associates, have demonstrated an interdependence between folic acid, vitamin B_{12}, and vitamin C in regulating the metabolism of phenylalanine and tyrosine. An interrelationship between these three vitamins has also been studied in chicks by Elvehjem and his coworkers who suggest a function of vitamins B_{12} and C in stimulating folic acid synthesis.

Darby and coworkers have shown that either folic acid or vitamin C will probable the abnormal tyrosine metabolism of the scorbutic gunieapig.

Rodney made research et al. with liver tissue of folic acid-deficient rats point out that only folic acid will probable the metabolic defect *in vitro* but that *in vivo* either folic acid or vitamin C is effective.

Vitamin B_{12} appeard to be both the much-sought antipernicious anemia factor and an necessary nutrient for many species, most likely including man. If its indispensability to man less clearly than to some other species, this is approach to be since of the reality that under favourable normal conditions a part of the human adult's requirement is given by the activities of the "friendly" bacteria in the intestinal tract; and also that infants may be born with prenatal stores from their mothers.

Vitamin B_{12} : Its Function

Vitamin B_{12} functions as a coenzyme in the metabolism of the body cells, particularly for cells of the bone marrow, the nervous system, and the gastrointestinal tract. Its role in metabolism is not yet properly understood. In the past the function of vitamin B_{12} had been concerned to the effectiveness in relieving the abnormal blood pattern and neurologic symptoms of pernicious anemia, due to its initial utilize in the relief of this disorder. From the hematologic viewpoint, vitamin B_{12} seems to be a factor in absence of which the reduced folic acid coenzymes cannot be formed or function effectively to assist in the production of normal red blood cells in the bone marrow.

The existence of the B_{12} coenzymes was first reported by barker in 1958. Since then data have accumulated on the biological role of these factors, and at the present time the B_{12} coenzymes have been concerned to seven dissimilar enzymatic reactions or processes.

Just after the invention of vitamin B_{12}, it was suggested that this nutrient might function to eveourage growth in animals and man. Supposing that a few isolated studies have been reported on the growth-promoting influence of vitamin B_{12} in small numbers of underweight children, further work is essential privious the role of the vitamin as a human development stimulant can be established.

Distribution of Vitamin B_{12} in Foods

Methods of assay for vitamin B_{12} are based on its growth-promoting influence for sspecific bacteria, for the rat, or the chick. Liberation of the vitamin and its separation from interfering substances, as well as the formulation of suitable medium, have led to some types of uncertainties in the consequences of microbial assay, but with improvements in technique and with corroborative consequences by the growth of rats treated with iodinated case in it seems specific that vitamin B_{12} is limited in it distribution in general foods. Scheid and Schweigert report values of 40.9-62.3 mcg. per 100 gm. fresh liver, 14.8-23.1 mcg. per 100 gm. fresh kidney, and from 0.9 to 3.2 mcg. for 100 gm. fresh muscle meat. Recent work connotes that apparent vitamin B_{12} values reported for alfalfa and other plant products may in part be since of some unidentified factor or factors, and that the richest food ways of vitamin B_{12} hence far tested are liver and kidney, with eggs, milk, and muscle meats contributing ample amounts. Robbins, et al. have suggested that the vitamin B_{12} values they find for root vegetables are due to the absorption of vitamin B_{12} formed by soil bacteria.

Vitamin B_{12} and Pernicious Anemia

Pernicious anemia is a chronic disease of genetic origin characterized by a macrocytic anemia, glossitis, and neurological disturbance. Stiffness of the limbs, irritability, drowsiness, and expression are some of the typical neurological symptoms. When liver was effective in treatment of the disease, Castle demanded that a materials in food (extrinsic factor) combined with a materials in the gastric juice (intrinsic factor) to form the antipernicious anemia factor. Vitamin B_{12} is now detected as the extrinsic factor. The intrinsic factor is a mucoprotein formed in the gastric juice. Pernicious anemia patients lack the intrinsic factor that functions in the body to remove vitamin B_{12} from the material to which it is bound and with the support of calcium to attach vitamin B_{12} to the intestinal wall. Thereafter, by enzymatic action, the intrinsic factor facilitates the passage of vitamin B_{12} through the mucosal cells into the blood stream. Very minute doses of the vitamin, 10 to 15 ug per day provided by injection (intramuscularly), are effective in relieving the disorder and substantial amounts provided every three or four weeks, prohibit the recurrence of pernicious anemia in patients (Fig-1).

It is mentionable that the observations of Herbert and Sullivan (10) on the relative effectiveness of vitamin B_{12} (cyanocobalamin) and coenzyme B_{12} in the treatment of pernicious anemia in a limited number of subjects. In the case of the deficiency with only hematologic damage, small doses of coenzyme B_{12} appear to be more effective in relieving the blood signs than do equal portions of cyanocobalamin. However, when both hematologic and neurologic symptoms were present, cyanocobalamin readily than coenzyme B_{12} was more effective in relieving the blood symptoms. From these limited data, these workers speculate that "... supposing that coenzyme B_{12} appears more potent than cyanocobalamin when only the hematopoetic system is damaged, a very small dose of coenzyme B_{12} may go preferentially to the nervous system when the system is destroyed, and hence be less hematopoietically potent in patients with both hematologic and neurologic damage."

Deficiency of Vitamin B_{12}

In 1956 the first documented case of pernicious anemia in man caused by an actual lack of dietary vitamin B_{12} was reported. The characteristic blood pattern was found in this patient who had existed on an animal-food-free diet for about eight years. The administration of vitamin B_{12}, however, brought about complete relief from the anemia. Deficiency manifestations may present in persons infested by the fish tape worm/ sprue, and other malabsorption syndromes.

Suggested Permission for Vitamin B_{12}

The human allowance for vitamin B_{12}, under normal conditions of health, is unknown, but it has been advised that a diet containing 3 to 5 ug per day should satisfy man's need. A diet comprising sufficient amounts of meat, eggs, and milk will above satisfy this need.

In a study by Chung and associates of high-cost, low-cost, and poor diets, the daily vitamin B_{12} contents were 31.6, 16.0, and 2.7 ug respectively.

Vitamin B_{12}: Its Sources

Animal proteins are the main way of vitamin B_{12} in the food intake. A summary of several assay reports of the vitamin B_{12} value of foods shows liver, kidney, oysters, and clams to be excellent ways of the nutrient. Lean beef, milk, veal, lamp, poultry meat, and salt water fish are believed to be good vitamin B_{12} foods where as cereal grains, vegetables, and legumes are poor carriers of the vitamin.

Vitamin B_{12} Folic Acid, and Choline

Liver extract has been found effective in promoting development and preventing the formation of fatty liver in rats on diets devoid of known methyl donors. In conformity with Bennet that folic acid was not the active principle, nor could the methionine or choline content of the extracts account for the effect. Following the isolation of vitamin B_{12}, a sparing influence of this vitamin on the choline requirement of chicks and rats was observed by Schaefer, Bennett and others. Jukes, finding out the influence of vitamin B_{12} on the response of chicks to homocystino, come to conclusison that vitamin B_{12} is included in the synthesis of methionine from homocystine. This was confirmed by Bennett and coworkers in 1951.

Folic Acid and Vitamin B_{12} : Its Nutritional

Function

The quantities of folic acid and of vitamin B_{12} needed in normal nutrition show readily large and as yet little-studied individual variation.

Folic acid occurs in both free and bound forms, and as tri- and hepta- (as well as mono-) glutamate. Additional forms of folic acid include an extra carbon atom linked as a formaldehyde or alcohol group. Conspicuously all of these forms are about equally present in animal and human feeding tests, but they vary widely in quantitative test with micro-organisns.

Both in the case of folic acid and of vitamin B_{12}, the output is more largely through the feces than through the urine, and little is yet known as to what part is due to the food and what part to bio-synthesis in the alimentary tract.

It is of greater significance to compare the graded responses to stepwise enrichment of diet by folic acid, with the consequences of graded intake levels of ascorbic acid as found by CG. King.

Register and Sarett studied ,the urinary excretion of vitamin B_{12} folic acid, and citrovorum factor in healthy young men on dissimilar diets, and found little influence of diet. In fasting the vitamin B_{12} excretion, but not that of the folic acid or citrovorum factor, was about double the normal value.

R.J. Williams and coworkers are of the view that the folic acid intake level of 25 mcg. per 100 grams of food suffices for survival in the chick; while 35 mcg. support normal hemoglobin formation; 45 mcg. support normal growth; 55 mcg. support normal feathering and 50 to 110 meg. support optimal growth.

11

Pantothenic Acid : Its Nutrition

Pantothenic acid became very famous in 1940 due to its possible utilize as agreemedy to over come the hair graying in men and women. It was observed that when the fur of a black-haired rat is deprived of vitamin it would grow grary. Based on the observation that the fur of a black-haired rat would gray when it was deprived of the vitamin, studies were undertaken to observe the impression of sufficient intakes of pantothenic acid on individuals with prematurely graying hair and on elderly people with gray hair. The findings from these studies connoted, however, that the vitamin had no consistent influence on relieving the process of graying hair in man.

This too, is one of the factors necessary for the development of yeast cells and known now to be required also by specific bacteria.

Historical Backdrop

Woolley and others, In 1939 it was syntheised, and

calcium pantothenate is now available as a pharmaceutical preparation.

In 1933 R.J. Williams coined the name *pantothenic acid* to connote global occurrence of this materials in all organisms which have been properly examined. It is most likely due to the wide natural distribution of this materials that there is as yet no consensus as to whether it require to be supplied to human beings through their food, or is formed adequately in our bodies or the micro-organisms to which our bodies serve as hosts.

Like many other vitamins, pantothenic acid is a essential component of an enzyme system, in this case, of a system controlling acetylation processes (such as the formation of acetyl choline) in the body.

Gopalan (1946) mentions the deficiency disease in his "burning feet." It has specific well-marked and constant clinical features and is a definite clinical entity. It seems to be general among the poor of South India-decidedly more general than "peripheral neuritis" concerned to thiamine deficiency.

Vernon (1950) evaluated and discussed a syndrome of painful "burning" sensation in the feet, observed in prisoners of war, which he approach had been frequently observed but "inadequately reported and referred to as dry beriberi." Vernon follows Glusman in calling the disease *nutritional melalgia* and further characterizes it as *a deficiency vascular disease.* Vernon writes, "It may represent a pantothenic acid deficiency," and in conformity with him, the work of Gopalan hints that calcium antothenate yields significant therapeutic consequences and "seems to establish a relationship of pantothenic acid" to the symptom of "burning feet."

"Outbreaks of 'burning feet' have previously been reported among malnourished populations and have often occurred in jails. The disease has not, however, hitherto been ssubstantially observed and evaluated and has often been confused with peripheral neuritis."

Signs suggestive of riboflavin deficiency were all but always available in the "burning feet" patients, but Gopalan considers the two diseases distinct.

Improvement on definite ground resulted in all of 10 patients who were provided calcium-pantothenate only. This improvement was more considerable and quick than that resulting from the yeast concentrate treatment which, however, was also effective.

The analysis for the close association signs of riboflavin deficiency with signs of pantothenic acid deficiency may lie in the reality that functions of these two vitamins in human nutrition are correlated and most likely in a similarity of their distribution in food. The observation of Spies and his associates (1940) that the injection of pantothenic acid reasons a development in the blood level not only of pantothenic acid but also of riboflavin, and correspondingly that the injection of riboflavin reasons a development in the blood level not only of riboflavin but also of pantothenic acid, is suggestive of a correlation of the two vitamins.

Pantothenic Acid : Its Functions

Pantothenic acid functions in the body as a part of the coenzyme known as coenzyme A. This coenzyme mediates acetylation, the transfer or acceptance of the acetyl group (CH_3CO). specific fats and amino acids can be formed from the intermediate products of carbohydrate metabolism by splitting off or adding an acetyl group (s). Coenzyme A is one of the enzymes mixed in the succession of chemical reactions that are must to break down carbohydrates and fats for the production of energy. The formation of the porphyrin part of the hemoglobin molecule needs coenzyme A. Pantothenic acid, a part of coenzyme A, is hence of importance in several phrases of cellular metabolism.

Pantothenic Acid : Its Sources

Inasmuch as this vitamin is found in all plant and animal tissues its name, which was coined from the Greek and means "from everywhere," is properly chosen. The richest amounts of the nutrient have been found in liver, kidney, yeast, egg yolk, and fresh vegetables. Like other B-vitamins, whole-grain cereals contain significantly more pantothenic acid than do refined cereal products. Milk, also, is a good carrier of the nutrient in the diet (1.5 mg pantothenic acid per pint).

Pantothenic Acid's Deficiency

In Man—The shortage of pantothenic acid in the diet, done, has not been effective in producing an experimental deficiency in man. Only when human subjects were fed an antagonist of the vitamin (omega-methyl pantothenate) in a diet deficient in other nutrients as well, were symptoms produced. Symptoms developed early in the experiment with the men showing signs of quick heart beat on exertion, dizziness, and a lowered blood pressure. Because they became fatigued easily, the men slept more in the day time than was normal. During the third week of vitamin deprivation, both anorexia and constipation developed, while in the following week the men became dissensive and discontented. As the deficiency progressed, the subjects complained of numbness and tingling in the hands and feet and displayed definite signs of muscular weakness. When pantothenic acid was restored to their diets, however, it alone did not soon improve all the symptoms of the deficiency state. Resultantly it is believed that the deficiency impression observed might have been due to the toxicity of the antagonist readily than to a complete lack of pantothenic acid; only the utilize of a diet supplemented with all the vitamins brought about complete recovery in all the subjects.

In Animals and Birds— The symptoms of a pantothenic acid deficiency in birds and animals vary from species to species. The chick displays development failure and then develops a characteristic sort of skin lesion which involves the eyelids and the mouth, whereas the fur of a black-haired rat will turn gray when the animal's ration is low or deficient in this nutrient (Fig.1). Other characteristic symptoms that have been observed when rats are deprived of the vitamin, are the failure to grow and the degeneration of the cells of the outer layer, the cortex, of the adrenal gland.

Suggested Permission for Pantothenic Acid

Although the daily allowance for pantothenic acid is Unknow hithey to it has been advised to be about 10 mg per day. The average amount in the substantial American diet has been reported to be between 16.3 mg (high-cost diet) and 14.2 mg (low-cost diet) per day and about 6 mg per day in insufficient diets.

12

Biotin : Its Nutrition

Biotin is a protective factor by which toxic influence of the avidin was counteracted. It knew by many name suchas factor X, Vitamins H.

Biotin was originally comprised in bios and was subsequent come to be known as vitamin H (from Haut = skin) since deficiency caused changes in the skin of rats. It was isolated in 1936 and synthesized in 1943 (Harris and others, 1943).

Biotin is present in each kind of living cell but in very low concentration. It is, however, one of the most highly biologically active substances known. The richest natural ways are liver and kidney, yeast and cereals.

Historical Backdrop

Since 1941 a sulfur-comp materials known as biotin has been considered as another of the B-group of vitamins. It first attracted attention as preventing the so-called "egg white injury" and has since been found to be essential in the nutrition of many species of animals. Conspicuouly the materials that reasons the injury since of excessive egg white is a protein called *avidin* or *avidalbumin*; which combines with biotin, formulating it unavailable to the body.

Biotin conspicuously has important functions in intermediary metabolism. It has been considered as a coenzyme in the synthesis of aspartic acid, which plays a part in a deaminase system; and in many other processes involving the fixation of carbon dioxide.

For further study see the advised readings at the end of the chapter.

It is still (1951-52) not sure whether biotin is apt to be of practical importance in human nutrition.

Biotin : Its Function

To date, the exact function of biotin in man is not known, but it appears that this nutrient is a neceary part of specific coenzymes. There is sufficent evidence that biotin is included in carbon dioxide fixation reactions and in the reverse of this reaction, decarboxylation. An intance of carbon dioxide fixation takes place in the conversion of pyruvic acid to oxaloacetic acid, an important step in carbohydrate metabolism. Carboxylation reactions are included in the synthesis of fatty acid and formation of urea. Biotin seems to be esential for biosynthesis of folic acid and is correlated metabolically to folic acid and pantothenic acid.

The Impression of a Biotin Deficiency

In Man—A biotin deficiency in man has been induced experimentally by feeding male subjects a low biotin diet that consisted of 30 per cent of the total Calories in the form of egg white. By the end of the first month on the experimental diet, all the subjects had developed a fine scaliness on their skin. In the ninth and tenth week of the deprivation, the men experienced mild depression, extreme weariness, sleepiness, pains in the muscles, and highly sensitive skin. Later, however, they developed anorexia and nausea. Since the symptoms of the deficiency state disappeared so rapidly after the restoration of biotin to the diets of the subjects, it was concluded that this vitamin is in essential nutrient for man.

In Animals—The symptoms of biotin deficiency in animals are readily resemble from species to species. In the early stages of biotin deficiency in rats, the animal loses hair around the eyes, which produces a characteristic "spectacle eye conditions," and as the disorder develops there is a common loss of hair over the entire body (Fig.1). Loss of weight, abnormal posture, and a spastic gait are observed in the advanced stages of the deficiency, and death will occur if biotin is not restored to the rat's ration.

Biotin : Its Sources

Biotin is available in many foods of both plant and animal origin. The richest food ways of the vitamin are found in kidney and liver, and good ways of biotin comprise eggs, most fresh vegetables, and some fruits. Meat, wheat products, and corn products, however, are said to be poor in biotin.

Suggested Permission for Biotin

As yet recommended allowances for biotin have not been established, but it has been advised that most likely between 0.150 and 0.300 mg of the vitamin give the daily needs. Because biotin is widely distributed in foods, with the average diet supplying the advised daily requirement more over being synthesized in the intestinal tract, it is very not likely that a deficiency of the nutrient occurs in man.

13

Choline : Its Nutrition

Historical Backdroop

Cholin is a methylated nitrogen-containing base. When sufficient amounts of amino acid, methionine are available cholin may be synthesized in the body. Choline is not a kind of vitamin. It occurs with vitamins of the B-complex in foods.

This is another materials which has been known for over a century but which has only recently been classed as a vitamin.

It was isolated from body fluids in 1849 and is known to be present in plasma, being closely concerned to another highly active physiological compound acetyl-choline. In fact, choline is, present in many animal and plant tissues, the best food ways being eggs, meat, cereals and some vegetables.

While differing from most of the substances which we call vitamins in that its chemical nature and wide natural occurrence were famous sevaral years privious its nutritional importance became explicit *choline* is now usually considered as a member of the "B-vitamin group."

It was noted by Best that choline prohibited the abnormal accumulation of fat otherwise occurring in the liver of depancreatized dogs or specific types of diet; and subsequent this same so-called *lipotropic* activity of choline was demonstrated in normal rats fed purified diets of high fat or cholesterol content.

The diets in some of the early experiments were so distorted as to leave significant doubt whether choline require be considered a factor in normal nutrition; but with increasingly successful endeavours to eliminate choline from the experimental rations, it became possible to produce symptoms of choline deficiency in absence of resort to grossly unbalanced diets. One of these symptoms is the disturbance of fat metabolism already indicated, which expresses itself in the growth of so-called fatty livers. But, as Griffith has displayed other tissues may be affected as well. Thus he found that on diets still more rigorously freed of choline than is must to induce fatty livers, young rats within a very short span of time showed "hemorrhagic enlargement and degeneration of the kidneys, a regression of the thymus, and an enlargement of the spleen"; and he advised that choline is necessary for the maintenance of the normal structure of tissues as well as for its lipotropic action.

These observations become very suggestive when it is recalled that choline is a component of some of the phospholipids which, as we have already seen, may have both structural and metabolic roles. The influence of choline in preventing the accumulation of fat in the liver was early interpreted as indicating that it facilitates lecithin synthesis, which, in conformity with one theory, is a significant step in fat metabolism. Direct evidence in favour of this view was adopted by Perlman and Chaik off who, using radioactive phosphorus as a tracer, showed that the rate of turnover of liver phospholipids is markedly accelerated by the administration of choline.

In sources not yet altogether clear, choline appears to be interrelated in its action with the sulfur-containing amino acids.

It appears also that choline may serve as a methylating agent in body processes. Thus it was found by du Vigneaud and his associates that [illegible]ocystine (the next higher homolog of cystine) failed to replace methionine in a diet rigorously freed of choline; but that it became a nutritionally satisfactory oprions for methionine when choline was also supplied. Their advised explanation was that a methyl group may be transferred from the nitrogen of choline to the sulfur of homocystine, converting the latter to methionine. The reverse transfer was also exhibited by the finding, after deuteromethionine (methionine containing deuterium in the methyl groups) had been fed, of a higher proportion of deuterium in the body of rats maintained on a methionine-choline-free basal diet. This observation may afford at least a partial analysis of the advised interrelationship of choline and the sulfur-containing amino acids.

Experimentally, choline has been found nutritionally essential to all species of animals with which it has been completely examined. The requirement appears to be comparatively greater in the young than in the adult, and to be increased in lactation.

In conformity with G. Emerson and K. Folkers identification of what appears to be a new member of the vitamin B complex, *L-lyxoflavin*. This new nutrient, a development factor for micro-organisms and rats, is reported to be similar to riboflavin except that the sugar L-lyxose is present instead of D-ribose.

Choline : Its Functions

Choline functions in protein and fat metabolism. It is a way of labile methyl (CH_3) groups which are utilised in the synthesis of vital protein substances like epinephrine, the hormone of the adrenal medulla. Choline is used in the formation of phospholipids like lecithin, a potent emulsifying substance. Choline clearly functions in another source be serving as the precursor of a concerned compound (acetyl choline) which is essential for the transmission of nerve impulses in the nervous system.

The Impression of a Choline Deficiency

In Animals and Birds. Not all species of animals and birds require dietary choline, but it has been shown to be an necessary factor in the ration of the rat, chick, and dog. A lack of choline in the diet of the rat consequences in a marked pooling of fat in the liver and an enlargement and hemorrhaging of the kidney. The fatty infiltration of the liver tissue and the kidney damage can be reversed with the administration of either cholinet or methionine to the animal. Vitamin B_{12}, however, has also been displayed to be effective in relieving the symptoms of this deficiency. Supposing that the characteristic influence of a choline deficiency in chicks is perosis or "slipped tendon disease," it is believed that both choline and manganese are essential for the prevention of this crippling influence in the chicken.

Suggested Permission of Choline

Because the impression of a choline deficiency have not as yet been found in man, no human allowances for the factor has been advised. The average American diet has been estimated to include 250 to 600 mg of choline.

Choline : Its Food Sources

The richest animal ways of this factor are found in egg yolk, glandular meats, and brain; wheat germ and the legumes are the richest plant ways of choline.

14

Inositol : Its Nutrition

Inositol

Inositol is a component of the body issue and not a particular nutrient is so since it is found in food stuffs and required by animals and micro-organisms in huge quantity.

This is another of the group of bios factors, required for the development of yeast. It is, such as biotin, available in all plant and animal tissues. The highest concentration in animal tissues is in muscle, brain, red blood cells, heart and kidneys. In plants it is found in yeast fresh fruits, vegetables and cereals, where, combined with phosphoric acid it exists as phytin.

The Impression of an Inositol Deficiency

Supposing that the function of inositol in the animal organism is not known, it has been observed that there is a halting of development when the factor is missing in the ration of the mouse, rat, guinea pig, and chick. Rats also show a loss of hair that is characterized by a "spectacle eye condition" suggestive of the lesion observed in a biotin deficiency of this species.

Suggested Permission for Inositol

Inasmuch as there is no evidence of the impression of inositol deprivation in man, the amount of this factor that one requires each day is not known.

Inositol Its Sources

Heart, liver, wheat germ, yeast, and all grain cereals have been found to be good food ways of inositol.

Para-aminobenzoic Acid (Paba)

This has been known as a constituent of plant and animal cells for about 100 years, but it was only in 1941 that it was detected as a vitamin in the B-complex group, being needed for development by chicks and for normal lactation and hair pigmentation of rats. It is also required by bacteria. It is present in most tissues but the best ways of supply are yeast, liver, wheat germ and rice polishings.

The Citrovorum Factor

In 1948, the existence of this materials was reported by Sauberlich and Baumann who found that *Leuconostoc citrovorum* — a non-pathogenic bacterium usually found in milk and dairy products—failed to grow in media comprising all previously known B vitamins and amino acids unless specific crude materials, like liver concentrates used in the treatment of pernicious anemia, were included. This unidentified materials could be partially replaced by high levels of folic acid and by high levels of thymidine. Following its occurrence in liver extracts and the low activity of folic acid, it was at first approach that there must be a metabolic relationship of the materials to vitamin B_{12} but it was just established by Sauberlich and Baumann (1949) and other investigators that the citrovorum factor (CF) differed from vitamin B_{12} and that the latter had no growth-promoting action for *L. citrovorum*. Sauberlich reported further that concentrates of CF overcame the toxicity of aminopterin (an inhibitory similar of folic acid) for *L.*

citrovorum, and that whereas rats on a folic acid-deficient diet excreted very little CF in the urine, administration of folic acid either with the diet or by injection resulted in large increases in the CF content of the urine of both rats and man. These consequences pointed to transformation of folic acid *in vivo*. Such a conversion in rat liver slices was reported by Nichol and Welch (1950), who also found that addition of ascorbic acid with folic acid greatly favored the conversion.

Researches of the activity of liver extracts in counteracting the inhibitory action of anti-vitamins concerned to folic acid (e.g., methylfolic acid) by Bardos, et al., and Bond, et al. gave advise a relationship of CF to formylfolic acid, a synthetic compound, which had been reported by Gordon, et al. to be more effective than folic acid in counteracting the prohibitory action of methylfolic acid for the organism.

Streptococcus faecalis—This was confirmed by Shive, et al., who reported the production from formylfolic acid of a mixture of compounds which was very active in promoting the development of both *Lactobacillus casei* and *Leuconostoc citrovorum* 8081.

Next came the isolation of a crystalline compound possessing the high activity of this reaction mixture as a barium salt by Brockman, et al., and as the free acid by Flynn and coworkers. Later work by Pohland, Flynn, Jones and Shive (1951) on the mixture of this acid led them to propose for it the tentative structure of 5-formy 1-5, 6, 7, 8-tetrahydropteroylglutamic acid. It is usually spoken of as folinic acid—SF (synthetic factor). That this synthetic factor is not identical with the natural factor has been established by Keresztesy and Silverman who reported the isolation from horse liver of the crystalline barium salt of the citrovorum factor. Assayed with *L. citrovorum*, this barium salt was found to be about twice as active as the synthetic factor.

Studies in higher animals with the synthetic factor have provided results readily corresponding to those obtained in micro-organisms. For insstance Brockman, et al. (1950) and

Broquist, et al. (1950) have reported that in mice 15 to 30 micrograms of SF overcame the toxic impression of 10 mcg. of aminopterin; in contrast, 20 mcg. of folic acid had no effect. Very much higher doses of folic acid-up to 5 mg. — were, however, found to be effective by Erschoff and coworkers

Both citrovorum factor (as a concentrate) and folinic acid (as the crystalline synthetic material) have been exhibited to have anti-anemic properties for man. The consequences due to intramuscular injections in pernicious anemia patients are less striking than those obtained with vitamin B_{12} or folic acid, but improvement in the blood picture, come back to normal of the bone marrow and improvement in clinical condition, but no modify in neurological status, have been reported by diverse independent groups.

Because CF and folic acid have similar activity for all organisms hence far used for the assay of folic acid, it follows that assays for the folk acid content of natural substance necessarily represent the CF plus folic acid content. What proportion of the total activity is due to each form is not as yet known. Schwarz (1951) has reported that in liver 95 per cent or more appears to be in the form of CF.

All studies hence far connote that folic acid must be transformed to CF within the living organism privious folic acid can carry out the catalytic functions for which it is required. In other words, the citrovorum factor either represents the catalytically active form of folic acid or is more closely concerned to this form that the folic acid itself.

15

Fat-Soluble Vitamins : Its Nutrition

Vitamin E

α (Alpha), β (Beta) γ (Gama) and δ (Delta) the four known natural to copherols are intimetely concerned that they are singularly known as vitamin E. Research that ultimatdy led to the invention of vitamin E was conducted in the early 1920's by Mattill, Evans, and Sure. In the process of fresearch of the ascendancy of diet on the reproductive cycle of the rat, it was found that rats became sterile when reared on a purified diet which contained all the then-known nutrients. Even though the experimental animals fed with this ration appeared to be quite healthy, they were not able to produce young. When fresh green leaves or dried alfalfa were fed to these sterile animals, reproductivity was restored in the female but not in the male of the species. The unknown factor, which prevented sterility, was first called materials X and subsequent was named vitamin E.

Synthetic alpha-tocopherol acetate has been followed as the International standard for vitamin E; and the International Unit has been defined as the certain activity of 1 milligram of the standard preparation. Vitamin E is readily widely distributed in nature where it is largely concerned to anti-oxidants.

Vitamin E : Its Properties

The chemical name for vitamin E is tocopherol (from Greek *tokos*, childbirth; *perhos*, to bear; and the chemical suffix -ol, signifying an alcohol). Indeed there are six dissimilar tocopherols, which in collective form are called vitamin E. The biological activity of the tocopherols varies, with alpha being the most potent. The number and position of the methyl (CH_3) groups within the molecule appear to ascendancy biological activity. This final opinion is arrived since the only difference among the tocopherols is in the structure of the molecule .

As a yellow viscous Vitamin E is find out that is insoluble in water but soluble in all the fat solvents. Supposing that it is readily stable to acids and heat (in the absence of oxygen), it is easily destroyed by ultraviolet light, alkalies, and oxygen.

Though Vitamin E is stable to heat but is destroyed when fats or oils containing it become rancid. It is hence safe from destruction as long as it remains in the intact cells but when the seeds are crushed or the oil is expressed, rancidity may develop with loss of vitamin activity. Vitamin E is also easily oxidized by ferric salts.

Vitamin E : Its Sources

The main vegetable oils which contain vitamin E are wheat germ oil, cotton seed oil and rice germ oil. A little also occurs in the small quantities of oil present in some particular green leaves, e.g. lecttuce and alfalfa.

In 1936, by Evans it was isolated in crysstalline form and since then various varieties have been identified, i.e. a-, B-, y - and 8-tocopherol. Mixtures of these appear to exist in natural ways in a state of equilibrium, but some plant products are richer in one and some in another. a~tocopherol, which is most active biologically has also been synthesized.

Vitamin E is extensively distributed in foods but not in high concentration. It is contained by bread it if made from the all grains but not if the germ has been removed. The animal foods containing the bigest quantities are eggs and lean meat—the latter since vitamin E is found in the animal body in association with muscle.

Inasmuch as vitamin E is found primarily in plant materials, the richest ways of the vitamin are found in vegetable oils (such as wheat germ oil and cotton seed oil), leafy-green plants and vegetables, as well as whole-grain cereals. Supposing that animal products contain little of the vitamin, liver, heart, kidney, milk, and eggs are the best animal ways of vitamin E.

Vitamin E : Its Function

For a very longer pat there has been a belief that the function of vitamin E in the animal organism is directly concerned to its antioxidant property. More over its possible function as an activator in particular enzymatic reactions, vitamin E plays an important role in the protection of vitamin A, carotene, and ascorbic acid from oxidation both in the digestive tract and in the body cells. Following the protective property of vitamin E more efficient utilize can be made by the body of vitamin A and ascorbic acid.

Chemically, this association with anti-oxidants in natural occurrence correlates with the reality that vitamin E, while relatively stable in many respects, is readily easily destroyed by oxidation.

The oxidative nutrition of vitamin E is catalyzed by iron salts. Hence a food mixture for experimental purposes can be made deficient in vitamin E by moistening with a solution of iron salt and then heating to dryness.

When rats are fed diets otherwise enough to their requirements but lacking vitamin E, the males become permanently infertile through an irreversible degeneration of the germinal epithelium; females fail in reproduction, the embryos being resorbed, but the female reproductive system does not suffer permanent injury.

The impression of dearth of vitamin E are not, however, restrained to the reproductive organs. The symptoms most prominent vary from species to species, and with the age of the individual, but in one or another species may comprise failure of growth, muscular dystrophy, injury to the central nervous system, interference with normal heart action, and feassible endoctrine disturbances.

Whether the counterpart of any of these conditions are found in human beings is still controversial.

Evans (1939) pointed out the occurrence of degeneration of the cross-striated musculature in vitamin E deficiency. It has also been advised that this may be the reaon behind the failure of growth of the fetus. Compare Mason (1944).

Puppies, rabbits, guineapigs, and rats have developed muscular dystrophy when kept for long periods on vitamin-E deficient diets, and in few cases this has been cured by offering synthetic alphatocopherol.

Vitamin E has also been found to have in some sense a sparing influence upon vitamin A. The converse may 'also be true.

Vitamin E occurs substantially in wheat germ oil, but may not be its only influential constituent as has been assumed by some experimenters. Other vegetable oils also are rich ways of vitamin E.

Following its comprehensive distribution in foods and the improbability of its being a limiting factor in human nutrition, it seems most likely unnecessary and probably misleading to lay emphasis upon vitamin E in practical considerations of food values. Though, there are diverse reported cases in which vitamin E seemed to be helpful in the treatment of specific muscular and nervous diseases but it should be noted carefully that the Council on Pharmacy and Chemistry of the American Medical Association has warned against in terms of the value of vitamin E for man as established, and has insisted the requare of more completely controlled investigation.

The impression of Vitamin E Deficiency

In Man—As yet there is no ultimate evidence to point out that vitamin E is necessary to man. It should be noted, however, that the plasma level of alpha tocopherol is lowered in normal adults after maintenance on a limited intake (2 mg daily in contrast to an estimated average of 15 mg daily for the supplemental group) for a period from 10 to 22 months. No clinical or physiological impression accompanied the lowered plasma level in the period of observation. Vitamin E seems to have no influence on the relief of sterility in human beings or on the cure of human muscular dystrophy. Hornit say that a relationship between serum levels of vitamin E and hemolysis of the red cells (erythrocytes) in the presence of hydrogen peroxide. Subjects who received the smallest amounts of vitamin E showed the biggest degree of red cell hemolysis.

In Animals and Birds—Vitamin E has been displayed to be essential for animals and birds, including the rat, rabbit, dog, guinea pig, chick, and duck. The most characteristic deficiency symptoms, however, occur in the rat and in the rabbit.

When rats are fed rations deficient in vitamin E, permanent sterlity in the male and temporary sterility in the female result. The permanent sterility of the male rat, that is caused due to degeneration of the cellular material of the testes, cannot be relieved by vitamin E therapy. On the contrany conception does occur in the deficient female rat, but about the eighth day of pregnancy the fetuses, or unborn animals, die and are resorbed. However, if vitamin E is added to the rations of these sterile females, their next pregnancy will be normal and they will produce living, healthy young. One of the causes postulated for the death of the unborn young is the dearth of an sufficient blood supply to the fetus due to a modify in the permeability of the reproductive tissues.

Young rabbits develop a condition known as muscular dystrophy when they are fed a ration deficient in vitamin E. This disorder, that is characterized by a loss of muscle tone and a general weakening of the muscles, is rapidly relieved when vitamin E is added back to the ration.

Suggested Permission for Vitamin E

In conformity with the food and Nutrition Board until more is known about the antioxidant needs of the body and the role that selenium and other nutrients play in decreasing vitamin E needs, it is problem to make any recommendation other than the vitamin E requirement will vary between 10 and 30 mg daily for adults. As vitamin E is so extenitively distributed in the foods of the diet, it is questionable that a vitamin E deficiency would ever develop in the human being. It is estimated that the average daily adult diet comprises about 14 mg of vitamin E.

Horwitt and associates are conducting a long-term study of the human requirement for tocopherol. Some of the conclusions from their work are—

(1) The tocopherol requare is a function of the amounts of polyunsaturated lipids in the diet and in the tissues.

(2) When polyunsaturated fatty acids in the diet are low, the requirement for tocopherol decreases to very low levels, but past dietary habits which have affected tissue composition must be taken into account in evaluating tocopherol requirements.

(3) The time of erythrocyte survival is shortened in man when a diet with a comparatively low tocopherol to linoleic acid ratio is fed for prolonged periods.

Vitamin K

A Dinish scientist Dam, in 1929 researched that chicks raised on a synthetic diet developed hemorrhages under the skin that could not be relieved with the addition of ascorbic acid to the ration. This disorder, which was characterized by a prolonged blood clotting time, disappeared when a mixture of cereals or natural foods was provided to the birds. Later Dam found that this antihemorrhagic factor was in the fat-soluble fraction of specific foods. He gave suggestion it be called vitamin K, derived from the Danish term "Koagulation Faktor."

Vitamin K : Its Sources

This is found naturally in green leaves, especially those of leguminous plants such as lucerne and alfalfa — also in similar an unrelated materials as fish meal. It was first illustrated in 1935 and was synthesized in 1939-40. specific chemically concerned compounds (analogues), which can be yielded in the laboratory have the same biological activity and these are the compounds primarily utilied in medicine. The fat-soluble analogue is menaphthone, and there is also a water-soluble compound menaphthone bisulphite.

Vitamin K is found in plants; good ways are alfalfa, cauliflower, cabbage, spinach, kale, and soybean. In contrast, fruits, cereals, and animal products include little vitamin K. The occurrence of vitamin K in a wide variety of commonly eaten plant foods, its synthesis by bacteria in the intestinal tract, its stability, and its insolubility in water make a deficiency of this vitamin in the normal person readily imposssible.

Vitamin K : Its Properties

Two forms of vitamin K are found in nature: K_1 occurs in green leaves, and K_2 is yielded by bacterial synthesis. These yellow-coloured vitamins are stable to heat, unstable to alkalies, strong acids, oxidation, and light. Many structural modifications of vitamin K have been synthesized; one water-soluble form, menadione, is more potent and more extensively used than the natural vitamin K.

Vitamin K : Its Function

Vitamin K is must for the synthesis of prothrombin which is a precursor of thrombin, one of the factors needed for normal blood coagulation. Even though the production of prothrombin explicitly occurs in the liver, the exact function of the vitamin in this synthesis is not known. It has been suggested, however, that vitamin K may be an essential part of the enzyme system included in the production of this blood clotting factor.

The impression of Vitamin K Deficiency

In Man—Vitamin K deficiencies in man result usually from either a faulty absorption of the vitamin or liver disorders that interfere with the synthesis of prothrombin, readily than from a dietary dearth of the nutrient. Inasmuch as vitamin K is supplied to the body by the food intake plus the synthesis by microorganisms in the intestinal tract, there usually is an sufficient supply of the vitamin available to the body. Sometimes however, the intake of larg amounts of sulfa drugs or other antibiotics may destroy the microorganisms which synthesize the vitamin and the supply then may become low.

As vitamin K is fat soluble, the presence of bile is must for its absorption as is true with the absorption of fats and other fat-soluble vitamins. Therefore, when disorders of the liver or gall bladder interfere with the secretion of the bile fluid, a vitamin K deficiency may occur. For instance, in cases of obstructive jaundice, where a blockage of the bile duct prevents bile from flowing normally into the intestine, vitamin K is not substantially absorbed. On the other hand, in particular liver disorders, like cirrhosis, ufficient vitamin K may be absorbed, but the synthesis of prothrombin does not occur since of the damage to the liver cells themselves.

Newborn infants may have inadequate vitamin K supplies since microorganisms of the correct type may not have had time to become established in the intestinal tract.

In Animals and Birds—Chicks fed a ration deficient in vitamin K develop a fatal bleeding disorder which is characterized by a longer clotting time of the blood. This disorder, however, may be relieved by the administration of vitamin K. It is a general faith believed that vitamin K deficiency is an avian characteristic. In some studies with rats maintained on a vitamin-K-free diet, when approach to their faces was entirely prevented, a bleeding tendency developed. Some vitamin K is synthesized in the small intestine by bacteria.

Suggested Permission for Vitamin K

The Food and Nutrition Board states that "with the exception of newborn infants, the average diet plus synthesis by intestinal bacteria give ample amounts of vitamin K."

The newborn child has low blood serum levels of prothrombin, which decrease further during the first week of life, but spontaneously come back toward the adult level in later weeks. As a consequence, it has been a fairly general practice to routinely administer vitamin K to the mother prior to delivery or to the new born shortly after delivery. The practice has been ubfect of controversy — it has been problem to demonstrate that decreased levels of coagulation factors are a direct reason of hemorrhage of the newborn (neonatal), and there is not total agreement that routine administration of vitamin K in pregnancy affects the incidence.

Vitamins K_1 and K_2

Danih investigation Dam, as early as, 1929, had observed that chickens raised on specific artificial diets were subject to subcutaneous and intramuscular hemorrhages; and further study of similar a hemorrhagic chick showed an abnormally long time required for the clotting of the blood. Others confirmed these observations; and Dam (1935) extended his work to the finding of experimental evidence that the hemorrhagic tendency and abnormally prolonged clotting time were attributable to scarcity of a fat-soluble "vitamin K."

The work of variou investigators showed that similar a factor is widely distributed in green leaves.

One form of antihemorrhagic vitamin, obtained from alfalfa leaf meal and designated as vitamin K_1 or phyllochinon (phylloquinone), has been detected chemically as 2-methyl-3-phytyl-1, 4-naphthoquinone.

A second form of antihemorrhagic vitamin (vitamin K_2) was isolated by Doisy and co-workers from putrefied sardine meal.

Apart from these forms of vitamin K known to occur naturally, many synthetic products show antihemorrhagic properties in greater or lesser degree.

Bile salts appear to be very important if not absolutely esential to the absorption of (natural) vitamin K. For this reason, patients with diseases of the biliary tract in which bile flow to the intestine is impaired are apt to develop a condition of low blood-clotting ability which is really a K-avitaminosis, despite the reality that their diet may have contained normally ample amounts of antihemorrhagic factor. Numerous clinical tests show that the hemorrhagic tendency in similar conditions may be controlled either by the oral administration of bile salts with vitamin K concentrates or by the injection of antihemorrhagic vitamin.

The low clotting-power of the blood in vitamin K deficiency has been traced to an abnormally low content of *prothrombin* (a normal protein constituent of the blood which, with calcium ion and the cephalin-comprising thromboplastic factor, gives development to blood clot). Hypoprothrombinemia has been found clinically in dissimilar conditions of impaired digestive function and treated effectively in most instances as a K-avitaminosis. Since diminished clotting power due to low prothrombin content seems to be usual in the blood of newborn infants, clinicians often recommend administration of vitamin K either to the mother privious delivery or to the infant at birth.

Vitamin P

A number of experiments since 1936, have advised the existence of another factor apart from vitamin C in the juice and rind of the lemon. This was at first called citrin and was subsequent found to consist of several concerned compounds, chemically known as glucosides, among which the best-known is *hesperidin*. Another glucoside with similar activity is *rutin* found in buckwheat and in the flowers of elder, violet, tobacco and forsythia.

Vitamin P activity is concerned to the control of bleeding from capillaries but is not corresponding to the action of ascorbic acid. There is still some doubt about its true nature and it was recommended in 1950 (Committee on Biochemical Nomenclature) that the term vitamin P, should no longer be employed. However, pharmaceutical preparations are available and ar being examined for their value in treatment and on the resulte of these will eventually depend the status of 'vitamin P.'

□□□

16

Preservation of Food

Vitamins are concerned in the development of the body from its starting until it has reached full maturity and with the maintenance of its structure throughout life. They are not structural constituents in the way that calcium phosphate is of bone or protein is of muscle but they are concerned in the formation and maintenance of the whole body of the diverse organs and tissues.

It is important to recognize that the body, while retaining its recognition and its individual characteristics throughout life, is in a regular state of change. The outer layers of the skin are shed and cells of the deeper layers constantly renew the covering of the body. This is a process which we can all observe. The same kind of renewal is all the time taking place even with structures as hard and clearly permanent as bone.

The initial development and continued renewal of the bodily structures depends in the first instance on food taken. That is the source of supply. But although food comprising of proteins, fats, carbohydrates, mineral salts and water, and the body comprising of the same there is a vast difference between bread and butter, meat and vegetables and the human or animal body into which they are transformed. This transformation to be effective, required the essential vitamins and each one plays some part in the complicated process. Some have functions still only very imperfectly understood. Others we know more about.

The deficiency diseases through which vitamins were first discovered showed how lack of a single vitamin caused a disturbance of function and also structural change in organs or tissues peculiarly sensitive to that certain factor, e.g. the link between vitamin A and the skin, vitamins B and the nervous system, vitamin C and capillaries, vitamin D and bone, but as soon as observations and experiences were followed by experiments, it was found that in all these deficiencies development was affected. Vitamin A and Vitamin B were, in fact, known as growth factors for experimental animals.

The first investigations into the ways where by these factors actually functioned in the body and thus brought about the effects on development, were made on vitamin B_1 which was found to be inevitable at certain stages in the breakdown of carbohydrates and the production of energy.

Vitamins and Carbohydrates **:** Carbohydrates (starches and sugars) are basically fuel foods. They supply the day-to-day requirements of the body for the maintenance of body temperature and for the performance of all physiological processes, whether involving muscular activity, glandular secretions or even mental and nervous functions. Carbohydrates do not become part of the body structure except in so far as they are stored as reserves. Glycogen is the form in which storage occurs in muscle and liver and carbohydrate eaten in excess over the daily needs may be stored as body fat, but in common these foods are eaten to give energy and this is liberated when they are chemically broken down (or metabolized) to produce at the final stage of the process, carbon dioxide and water.

Diverse chemical stages are controlled by enzymes (compounds produced within the body, which accelerate or regulate chemical reactions). Lactic acid and pyruvic acid are intermediate products produced in the period of the breakdown of carbohydrates and in deficiency of vitamin B_1 is associated with the enzyme which is concerned in the oxidation of pyruvic acid. It is, in fact, a co-enzyme and is responsible for the control of chemical oxidative processes which occurs within the living cell in connection with its energy relations. Since vitamin B_1 performs such a important function as this, it is not surprising that shortage affects many physiological processes in addition to inhibiting growth.

Vitamin B_1 Functions: Being readily soluble in water, vitamin Bj is quickly absorbed from the digestive tract and enters the blood stream to be carried to the tissues of the body. It is utilised continuously in the important physiological processes and the tissues are not able to store it except for short periods. Consequently regular supplies are needed for health.

The effect on development is seen in children, who make only subnormal gains in weight if their food is short of vitamin B,. At the same time their appetite may be poor, the slackness of their intestinal muscles may cause constipation and they may lack the mental alertness of generally healthy children.

The same apathy and lethargy affect adults, abnormal fatigue may be due to of excess of such breakdown products as lactic acid and pyruvic acids in the tissues. All stages occur between such common signs of debility and the symptoms and signs characteristic of beriberi.

In this disease, the loss of appetite and consequent reduced food intake may produce extreme emaciation, but this may be masked by the presence of fluid in the tissues causing swelling of the limbs. The heart gets always affected in beriberi, being significantly enlarged and its rate is increased, so that when death occurs, this is frequently from failure of the circulatory mechanism.

The nervous system changes that causes abnormal sensations in the limbs, extreme weakness, wrist-drop and loss of knee-jerks.

Relations with other Vitamins : Riboflavin and nicotinic acid also participate in the same enzyme systems concerned in the intermediate breakdown of carbohydrates. It is feasible in experiments with animals to make diets deficient in one vitamin only and to study the effects of deficiency of one factor at a time but deficiency diseases in man rarely occur singly. The effects of one deficiency could be more marked than those of another but even in extreme cases like beriberi and pellagra, many factors are involved.

***Riboflavin* :** Although riboflavin is more concerned in carbohydrate oxidation, deficiency is shown by effects rather different from those caused by vitamin B, deficiency. The chief characteristics are cracking of the lips with sore, red patches, scaliness with raw cracks at the corners of the mouth and a sore, red tongue. There may also be scaly red patches across the nose and lips and over the eyelids and sometimes redness of the eyes. These signs may seem in individuals even in Great Britain if their diet has been poorly supplied with such riboflavin-containing foods as milk, eggs and liver.

***Nicotinic Acid* :** Nicotinic acid is also fundamentally a tissue co-enzyme and is inevitable, along with vitamins Bj and riboflavin for the breakdown of carbohydrate, but extreme deficiency of nicotinic acid causes not beriberi but pellagra, the disease that is still prevalent in countries where maize is the staple cereal. The word is derived from pelle=skin, agra=rough, and was utilised first to mentined this disease in 1771 in Italy. It was Funk, who had coined the word 'vitamine', that suggested in 1914 that pellagra was a deficiency disease. Its exact cause has given rise to much controversy but can possibly be explained as follows: Maize, like any other cereal grain, consists of both carbohydrate and protein and the protein is made up of a collection of simpler nitrogenous compounds, the amino-acids. Some of these are essential for life and health; others can be dispensed to be very poorly represented in maize. It has been known that the occurrence of pellagra was relateed to the consumption of foods poor in protein. The connection with nicotinic acid was recognized when it was found that tryptophan was a precursor of nicotinic acid. The disease disappears when food protein containing tryptophan or when nicotinic acid (or amide) is provided.

Pellagra is characterized by soreness of the tongue and mouth, presistent diarrhoea and the development of pigmentation of the skin leading to dermatitis. At the same time patients become mentally depressed. The pigmentation of the skin takes place in exposed parts : hands, face and neck.

The skin is dark and brown, later becoming red and scaling with raw patches. The depression may grow into a severe dementia.

Pellagra is Fairly Common in Parts of America and in the Middle East: Mild degrees of nicotinic acid deficiency is found in other countries if individuals fail to get their quota of nicotinic acid and protein foods, or if come other condition prohibits either absorption or utilise by the body of these food substances. The chief sources of nicotinic acid in human foods are meats and fish.

Hence when people eat only a little of these and large quantities of maize, they lack both the protective factors, tryptophan and nicotinic acid, and pellagra is the outcome. In animals other vitamins, and in particular &$_1$ and B_6 are needed for the full utilization of nicotinic acid and also for its formation from tryptophan - another instance of vitamin interactions.

B *Vitamins and Proteins* : Riboflavin and nicotinic acid are not only concerned with carbohydrates and energy relations but also with protein and the muscular structures of the body.

All proteins are made up of amino acids and there are some twenty of these latter, which, combined in diverse ways and in different proportions account for the characteristics of the varied proteins as we know them. For intance lean meat is protein but we know from our own experience that there are differences between beef and mutton and between mutton and pork; and even greater differences between meat and milk, cheese or eggs. But the body can use all these proteins as sources of the amino acids which are to be made into human muscle. Protein eaten in any shape is first digested and absorbed and then its amino acids have to be rearranged and built up again either for the development of the human body or for its maintenance when it is hilly grown.

In this process both riboflavin and nicotinic acid play a part. Moreover, riboflavin, when eaten in food, is transported to the liver where it is temporarily stored, to be released again when required in the period the digestion and assimilation of food. It has been shown that for this release from the liver vitamin B_x is required.

Thus these three vitamins are very intimately connected not only in foods where they are found in nature, but also in their functions in the animal body, and we can see why the absence of any one of them interferes with growth and with common well-being.

B *Vitamins and Fats* : Just as particular vitamins have a special function in the metabolism of carbohydrates and others in that of protein, so a few seem to have close association with the body's use of fats in relation with either absorption or subsequent use.

Fats are digested in the small intestine and afterwards pass through the cells lining the tract into the lacteals and thence to the Lymphatics.

After absorption, fat is carried first to the liver and thence passes to other parts of the body, being either broken down for the production of heat and energy like carbohydrates or laid down in varied fat depots for storage.

Choline : Choline is concerned in the utilises of fats by the body. It is needed first of all for their absorption through the cells lining the intestinal wall and it is required subsequently by the liver for the conversion of fats into substances suitable for passing in the circulating blood to diversse parts of the body. Fats do not circulate as such but as compounds with phosphorus, phospho-lipids, and part of the choline molecule is needed for this conversion. For this reason choline called a lipotropic agent and in this respect is similar to inositol and vitamin B_{12}.

Lack of choline affects the functions of the liver, but hi the absence of choline, an amino acid, methionine, can take on some of its functions. Due to this reason some doubt has been cast on the justification for considering choline a vitamin since it is not absolutely inevitable for life. However, now it is usual to comprise choline among the vitamins and its role in metabolism is certainly similar to that of other vitamins.

In the absence of choline, fat collects in the liver, interfering with the proper functioning of the hepatic cells.

This led to dyspepsia, and if the process continues, the liver may become permanently damaged. Since normal metabolism of fat depends on supplies of choline, animals on a choline-deficient diet grow badly and diverse signs of deficiency develop, for instance, in substantial lactation in the rat and slipped tendon in chicks.

Inositol: The function of inositol in the animal body is only partly understood. Such as choline, it assists in the metabolism of fats and it may have some effect on the skin. It has also been suggested that with vitamin E it affects the health of the muscles.

Inositol itself is soluble in water, but in plant products it is generally found as phytin, a compound with phosphorus, the calcium salt of which is insoluble. This fact gave rise to the view that cereals containing inositol caused rickets by prohibiting the absorption of calcium. It has been found however, that this only happens in severe deficiency of calcium and that with normal supplies of calcium and vitamin D, no apprehension require be felt on this score.

***B_6 or Pyridoxin* :** The role of this vitamin is not yet clearly understood Animals on a deficient diet develop eczema of the skin particularly of the feet, but nothing corresponding closely with this seems to occur in man. However, the vitamin has been found useful in medicine for the treatment of many conditions, e.g. nausea and vomiting in pregnancy and radiation sickness and is one insstance illustrating the fact that a vitamin may be found of value before an explanation is forthcoming of exactly how it works. One fact that is known at how is that it is associated in the body with protein metabolism.

***Biotin* :** Experiments with animals on restricted diets have demonstrated that this is essential for the development of almost every animal and also of yeasts, bacteria and fungi, but due to its almost universal presence in human foods, little is known of deficiency in man or of human needs.

We know, that raw egg white combines with it in the intestinal tract rendering it insoluble. The factor in egg white is liable for this action is referred to as avidin because of its avidity of biotin.

There is thus a very slight danger that individuals consuming excessive amounts of raw egg white may become biorin-deficient. Actually this is a rather remote feasiblility, but sick children in hospital, given abundant quantities of albumin water, have been reported as developing a form of scaly dermatitis that has cleared up with small doses of biotin. These are,, however, only isolated observations and have not been confirmed commonly.

Pantothenic Acid **:** Rats and chicks fed on diets deliberately made to comprise no pantothenic acid develop diverse signs of deficiency: e.g. dermatitis in chicks and development failure, hemorrhages and changes in the adrenal glands in rats.

Deficiency in man is not likely to occur since the vitamin is so widely distributed in all kinds of foods.

It has been suggested from time to time that pantothenic acid affected graying of the hair but this has not been confirmed. It has been utilised in medicine in doses higher than those which would normally be taken in foods. These comparatively high doses have been found effective in the treatment of several conditions comprising the sore tongue which some patients with pernicious anaemia develop. Pantothenic acid is not of any value in the treatment of the anaemia itself.

Para-amino-benzoic Acid: As a dietary factor this presents no problem since there is little risk of shortage in foods, but like several other vitamins, it has found a place in medicine given in high doses and has been used in the treatment of diverse forms of eczema and also of rheumatic fever.

Vitamin D and Bone **:** Among animals with a bony skeleton, the form, shape and size depend on the development and perfection of this skeletal framework. The bones are chemically of calcium phosphate, but each bone has its own characteristic architecture and internal structure. The bones begin as relatively soft cartilage and development the hard calcium phosphate is deposited in them, being carried there in the blood stream.

The formation of bone from the calcium and phosphorus brought by the blood is controlled by vitamin D. The body cannot make good bone (or teeth) without vitamin D and conversely the vitamin cannot produce good bone without the materials from which bone is made. Calcium and phosphorus, together with traces of other elements like fluorine are the building materials; vitamin D regulates the process.

The supply of these minerals and vitamins reaches its greatest importance in the the period of rapid growth and it is when the bones are young, cartilaginous and relatively soft that deficiencies may cause rickets with the characteristic bending of the bones of the limbs and widening at the ends. It is note worthy however, that since bones are hard, they are not of unchanging structure like a piece of wooden or metal furniture. Every bone is supplied with blood vessels and nerves and is as much a living part of the body as muscle or skin. Moreover the blood is regularly carrying new nutrients to the bones and removing waste products. There is thus a state of balance or equilibrium between the calcium and phosphorus of bones and of circulating blood and this continues whole life.

The amount of vitamin D needed during active growth is definitely greater than that required when the full size of the body is reached, but some vitamin D continues to be needed to maintain the bones in a healthy situation. Old people, whose consumption of calcium and vitamin D has been neglected, sometimes grow very brittle and fragile bones which fracture easily — and in some cases the bones of the spinal column lose their firm hard texture and collapse, causing great pain and weakness.

Vitamin D is the one vitamin that can be yielded by sunlight, vitamin D is formed in the oil on the surface of the skin and in its deeper layers. This is absorbed and in sunny climates makes it less essential for sources of vitamin D to be taken by mouth.

Vitamin D is also formed of the fur and feathers of mammals and birds and is consumed when they lick themselves or preen their feathers.

The fact that vitamin D is formed when the skin is exposed to sunlight has led to its use in high doses for particular diseases formerly treated by ultra-violet irradiation. One such disease is lupus vulgaris (tuberculosis of the skin) in which outcomes have been encouraging.

Bones and other Vitamins : Concentration on the importance of vitamin D in the growth of bones and teeth led to neglect of other factors possibly concerned. However, vitamins A, C and E all seem to play some part in promoting normal bone formation.

In vitamin A deficiency the structure becomes disorganized — the bone becomes cancellous (less hard and compact) and overgrown, losing its fine moulding and outline.

In infantile scurvy, the ends of the long bones are abnormally brittle and fractures repeated occur. Vitamin E deficiency is thought to lead to imperfect calcification of the skull, feasibly through its effect on vitamin A or on the phosphorus 'turn-over' in bone.

Art of Preservation

Vitamins A and the Skin: All parts of the body whether external or internal are covered or lined with skin. The kind of skin differs according to situation, being tougher and thicker on the outside of the body and more delicate when lining internal cavities, but carrying out important functions throughout the life of any animal.

Vitamin A has special functions in connection with the skin, both that covering the outside of the body and also the mucous membrane lining the different parts which communicate with the exterior, e.g. the mouth, nose, trachea and bronchial tubes, the conjunctiva of the eyes and the linings of the eyelids.

The eyes are peculiarly sensitive to severe deficiency and the situation of xerophthalmia (dry eyes) was one of the first signs of vitamin A deficiency to be discovered.

The mucous membranes of the respiratory tract become less healthy if there is a shortage of vitamin A and because of to this fact vitamin A was considered 'anti-ineffective.' We know now that it has no direct effect on infective organisms but it may help to prevent respiratory infections by maintaining the mucous membranes in a healthy situation. The outer skin of the body shows little change in mild degrees of deficiency but subjects suffering extreme deprivation for long periods may develop a horny, dry situation of the skin — keratomalacia.

Other Functions of Vitamin A **:** Much has been discovered about the connection between vitamin A and vision in dim light.

Deficiency causes a situation called night-blindness in which individuals experience difficulty in adjusting their sight to distinguish objects in a poor light especially after being in a bright light or looking at bright objects. Normal persons can adapt themselves readily to changes in light intensity but vitamin A-deficient individuals have poor dark-adaptation. It can be described as follows: seeing in semi-darkness relies on the presence in the retina of the eye of a pigment, visual purple. When bright light falls on the eye, the visual purple is bleached. If there *are* ample supplies of vitamin A in the body, the pigment is reformed, but if supplies are poor, this re-synthesis is delayed with the outcome that the clarity of vision in a dim light is reduced. Visual purple is actually a chemical derivative of vitamin A itself.

Above facts from the basis of tests used to find out the state nutrition of groups of people in connection with vitamin A. By means of it, subjects not otherwise unhealthy have been detected as in a state of mild deficiency and it has been possible to correct the faults in their diet before any serious consequences have grown.

***Vitamin E and Body Tissues* :** The function of vitamin E are less well deemed than those of vitamins A and D there is still much conflict of opinion about its role in human nutrition.

It is absorbed from the digestive tract along with fats, and anything which interferes with fat absorption may lessen the quantity reaching the tissues of the body.

In experimental animals, e.g. rats, it has been proved that vitamin E is essential for normal fertility and this fact is still used in biological methods of estimation, but because of this, it cannot be assumed that the same is true of other species.

Although the role of vitamin E in connection with fertility is still not deemed and appears to vary from species to species, there is no doubt at all about its significance in the physiology of some of the tissues. Just as vitamin A is related to with the skin and mucous membranes and vitamin D with bones, so vitamin E is related to muscle and connective tissues. Deficiency in man does not produce such well-marked and easily recognized diseases as xerophthalmia and rickets but in animals following rats, the young grow a paralysis of the limbs due to a muscle defect or dystrophy. This is not unlike a rather uncommon human disease but efforts to cure the latter with vitamin E have so far met with very inconclusive results.

One function of vitamin E that appears to be well established is its role as an 'anti-oxidant.' When present in the tissues or in foods with readily oxidizable constituents of fats, comprising such substances as vitamin A and carotene, it prevents their destruction by oxidation. It is thus stated to 'save' vitamin A. This effect can occur both in the intestinal tract where vitamin A in food may be oxidized, or in the liver where vitamin A is stored, and this fact means that if vitamin A is given *with* small quantities of vitamin E, less A is required to achieve a given result. It has been shown, for example, that vitamin E increases the growth-promoting power of vitamin A.

It may be partly due to this property and partly for other reasons not fully understood that vitamin E is as essential for the maintenance of tissue integrity and is thought to have an effect in delaying the process of ageing.

Vitamin E is one of the best instances to illustrate the complications with which a study of vitamins may be included. There is no simple deficiency here. The vitamin itself is made up of a complex of at least four components. Their biological activity differs; their distribution varies from source to source; they are known to interact inside the body with proteins and protein constituents, the amino acids, with vitamin A and perhaps vitamins of the B complex and to be relatied in the metabolism of some of the most complex tissues, the muscular and fibrous tissues, of the body. It is not surprising in the circumstances that there are still several gaps in our knowledge of this certain vitamin.

Vitamins, Blood and Blood-Vessels: Vitamins C, K, P, folic acid and B_{12} are all concerned either with blood or with the smallest blood-vessels, the capillaries, which bring blood into relation with all the tissues. Blood is made up of a fluid, plasma, of highly complicated composition where in are suspended cells or corpuscles: red cells containing the pigment hemoglobin and white cells which are unpigmented.

The blood, by circulating through the vessels comes into relation with all the tissues, even the hardest bone. Holes can be found in the shafts of the long bones and in the skull, through which these vessels pass to the interior.

As the fluid (plasma) circulates, it carries the cells with it and together they form the distributing and gathering system which functions for the whole body. Oxygen is carried from the lungs to all parts; carbon dioxide, formed in tissue respiration, is removed from its sites of formation and taken back to the lungs. For this, the red cells are very liable.

The plasma transports nutrient materials from the alimentary tract to the organs or tissues where they undergo further breakdown or assimilation. The waste products of physiological processes are similarly removed by the plasma, which may comprise at any one Lime proteins, inorganic and organic substances in solution, nutritive and excretory materials, antibodies, hormones, vitamins and feasibly compounds still unidentified.

By studying the effects of deficiencies of vitamins C, P and K, we can see how they affect the blood. All three are associated with bleeding into the tissues.

***Vitamins* C *and* P:** In scurvy, there are haemorrhages into the joints and into the tissues. These appear as purplish-red patches in dependent parts, i.e. around the shoulder blades and lower part of the back when the patient is lying down. The gums may also bleed, and there is swelling because of an enhance in the amount of fluid in the tissues. It is clear that the capillaries have allowed blood to escape.

The walls of these minute vessels are made up of a single layer of very delicate cells, held together by a cement substance and surrounded by a sheath of quite fine fibrils.

Bleeding into the tissues occur when blood-vessels of any size are damaged. This can happen in mechanical injuries and the severity varies according to the size of the vessels which are destroged. Bruising causes injury to the walls of the capillaries and the discolouration that follows is became of the presence of blood in the tissues.

When similar escape of blood takes place without mechanical injury' it means that the walls of the capillaries have become unduly permeable. This is what happens in scurvy and in another condition : purpura — a term applied to a number of situations in which spontaneous bleeding occurs beneath the skin. The capillary sheath and the cement substance holding the cells together give way and allow the blood to escape.

The haemorrhages seems as purplish spots of various sizes and the colour changes which take place are similar to those of an ordinary bruise.

Purpura occurs in diseases of diverse sorts — several quite unrelated to scurvy, and it should not be looked upon as a disease in itself but rather as a disease sign. In the most serious varieties of measles, scarlet fever and smallpox, purpuric spots seem and the diseases are then described as 'haemorragic'.

Purpura may also take place in situations of irritation of the digestive tract and related to weals in some kinds of urticaria (nettle-rash).

Scurvy can be arrived and treated by means of vitamin C; purpura by vitamin P.'

Vitamin C (ascorbic acid) is concerned not only with the prevention of tissue bleeding as in scurvy, but in the maintenance of a healthy state of equilibrium between the fluid plasma in the capillaries and the fluid in the body tissues (Hoskin, 1948). Where excess body fluid tends to gather as in diseases of the heart and circulation, vitamin C helps to reduce this.

Vitamin K: This is concerned not with the capillary walls but with the process of clotting, which is slowed down in deficiency of the vitamin. Clotting is a complex process. It does not generally occur inside the intact, healthy blood-vessels but when the vessel or the tissues are detroyed the blood which escapes, clots within a few minutes. The clot is made up of fibrin, a jelly-like protein that forms in the plasma and that entangles the blood cells in its meshes. This fibrin sets as a jelly only in the presence of another constituent of plasma, thrombin, and this in turn is formed from a precursor prothrombin, that is liberated into the circulating blood from the liver where it is formed. And this is where vitamin K comes into the picture. Without it, there is a failure in the maintenance of prothrombin in the liver and if the amount liberated into the plasma is much subsidied there develops a tendency for injuries, which would otherwise be negligible, to continue to bleed since no clot forms to seal them.

In the blood of infants, the prothrombin is low for the first few days of life and consequently hemorrhage may take place from undetectable small injuries to blood-vessels. These haemorrhages may be in the digestive tract when their existence is detected by the passing of discoloured stools, or in the nose or palate or under the skin, hence the descriptive term 'haemorragic disease of the. newborn.'

The low levels of prothrombin in the blood of infants may be due to the fact that vitamin K is received only to a small extent from food but chiefly as a result of bacterial synthesis in the digestive tract. Only relatively few bacterias are related to with disease; some well the digestive tracts of man and of other animals and their presence is an advantage rather than the reverse.

Although vitamin K is eaten with particular foods, there is no doubt but that much of a man's needs for this vitamin are supplied by these bacteria in the intestines. Vitamin K cannot, however, be synthesized by the intestinal bacteria until these are setup and this does not happen until many days after birth, so that during this short period there is a real danger of vitamin K deficiency and subsequent haemorrhage. To prohibit this, it is now a usual practice to give vitamin K either to the mother before the child is born or to the infant after birth and this has done much to reduce the incidence of this kind of haemorragic disease.

Vitamin K, whether formed by bacteria in the intestine or eaten in foods, is fat-soluble and is so absorbed along with fats. Fat-absorption, in turn, rehies on the presence of bile in the intestine since bile salts act on fats converting them into a form in which they will pass through the cells lining the intestine. Anything which intervenes with the presence of bile salts in the intestine will therefore prevent the absorption of fats and with them of vitamin K.

The commonest such situation is an obstruction in the bile duct, through which bile passes from the gall-bladder to the intestine. A gall-stone in this duct is a common form of obstruction and a construction from a growth can produce the same sort of result.

When the bile can no longer pass into the intestine it becomes re-absorbed into the blood stream and the pigments in it circulate in the blood and turn the skin yellow producing the situation known as obstructive jaundice.

Occasionally surgical operations are required to remove the obstruction and these operations formerly attended with some danger from haemorrhage since, with poor vitamin K absorption from the intestine, the liver had failed to produce substantial quantities of prothrombin. Now, vitamin K is given to such patients before operation and this danger is greatly reduced.

A secondary deficiency of vitamin K may occur in any disease where in there is defective absorption of fats. In fact individuals may have low prothrombin levels in the blood with no signs to show this until some minor damage to the tissues occur such as a tooth extraction, after which bleeding may be excessive. The practice of giving vitamin K before such minor operations is becoming more prevalent.

The other two vitamins in this group effect the cells of the blood rather than the plasma. Red blood cells are formed in the red marrow of bones and go through stages in growth before they are fully mature and ready to be released into the circulation. They remain in circulation carrying out their diverse functions for about four months after which they tend to become fragmented and are disposed of by the spleen and new ones take their place.

Primitive cells in the marrow comprise no haemoglobin and they stain with basic dyes, so that in stained films a cell seems blue or purple with a large nucleus. Steadily haemoglobin is formed inside the cell, and as this stains with acid dyes (usually pink) the cells appear in films with both pink and bluish colouring. These cells are larger than circulating red cells and are known as megaloblasts. They gain more haemoglobin, lose their nuclei and become smaller before being liberated into the circulation as normal red cells or normocytes.

In those cases where this process for some causes or other goes wrong, various things can happen; cells may be liberated with less than the normal quantity of haemoglobin, or before they are fully mature, or are of uneven size.

These abnormalities of blood cells or the amount of haemoglobin is reduced. All these are anaemias of that various kinds are known. Those types that are characterized by abnormally large red cells (macrocytes) and by a reduction of haemoglobin are the macrocytic anaemias, the most common of which are the so-called pernicious anaemia and the anaemias that occur in the disease, sprue. There are other forms of macrocytic anaemia.

Folic Acid and Vitamin B; It has never been proved that these anaemias are directly because of dietary deficiencies, but it is now known that a normal blood picture can be restored by both folic acid and vitamin B_{12} However, the mechanism by which this is brought about is still only vaguely understood.

The story of vitamin B_{12} has been, in recent past, as dramatic as the history of the earlier vitamin discoveries. For the starting we have to go back to 1926 when Minot and Murphy found that a form of anaemia hitherto resistant to treatment and hence described as 'pernicious' responded to treatment with liver. From that time onwards chemists sought to isolate the active factor from liver and, after about twenty years, the goal was thought to be reached when folic acid was synthesized. This substance restored the blood of pernicious anaemia patients to normal but this disease is characterized not solely by abnormal blood formation but also changes in the nervous system causing disturbances of sensation and these were not affected by folic acid.

In 1948, workers in the laboratories of pharmaceutical firms in the U.S.A. and in Great Britain simultaneously announced the isolation from liver of a crystalline compound many thousand times more active than folic acid. The crystals were dark red needles and contained, in addition to carbon, hydrogen, oxygen, nitrogen and phosphorus, 4 per cent of cobalt — the only vitamin known to comprise this element.

The new substance, called vitamin B_{12} was found to be active in pernicious anaemia in doses as small as 20 , micro-grammes and also effective in the treatment of the nervous system complications.

It was marketed in Great Britain in 1949 and is now widely used in doses of 50 to 100 micro-grammes given by injection. It is no longer derived from liver, twenty tons of which are need to produce one gramme, but is produced from cultures of streptomyces griseus (an organism allied to primitive moulds). It is stable in the solid state and also in neutral or slightly acid solution. It exists in many forms — all physiologically active. The solution, which is pale pink, is supplied in ampoules for medical use.

Goodness of Stuff

A large majority of our people are satisfied by filling their bellies with some foodstuffs. Most of the people reject the nutritious aspect of the food they eat. Prevalence of malnutrition is because of insufficient intake of foodstuffs. In order to assess the nutritional status dietary intake must be calculated. Diet surveys are used to gather information about the dietary intake, dietary habits, preservation methods, cooking methods and special diets for special situations and in diseases.- Diet surveys also give information with regard to the economic and social factors influencing food production and consumption patterns of families and communities.

To conduct diet surveys certain factors are to be taken into account. They are trained personnel, population sampling, diverse methods available for conducing diet surveys and analysis of the data and calculations of nutritive values.

The persons to be engaged in making diet survey must be trained in conducting such surveys. They must be trained to create rapport with the housewives and in adopting tactful methods to gather information from them. They must be aware of other principles of nutrition and dietary habits of the local community.

Population sampling must be carried out with the help of a statistician. The choice of diet survey methods rehies on the purpose of the survey.

The sample size also relies on the purpose of the survey, availability of trained personnel, time and resources like money. The duration of the survey relies on the purpose and methods followed. Normally 7 to 10 days are preferred for diet surveys.

Different Processes : Commonly utilized methods in our country are—

(1) oral questionnaire method,

(2) food inventory or log book method, and

(3) weighing of raw and cooked foods.

Oral Questionnaire Method: Oral questionnaire or interview is the most commonly used method of diet survey. The investigator has to formulate a diet survey questionnaire relying upon the purpose. After making a pilot study with the questionnaire changes can be made on the pattern of the questions. Then the investigator goes from door to door and gathers information regarding the types and qualitiès of foods consumed. The data is not very perfect, only approximate amounts can be gathered through this method. Therefore, this method is suitable only to gather information on the general dietary patterns or an approximate picture of dietary habits of a large section of population. General ideas like choice of foods, foods avoided in certain situations, foods included in special situations or in festivals, food fads and fallacies, are to be gathered. The advantages of the questionnaire method are that it is not a time-consuming method, therefore a large number of families can be covered inside a specific time.

Food Inventory or Log Book Method: This method can be used only with a literate group since a book with question is entrusted to the head of the family or to the housewife. The person entrusted must enter all purchases in the book. Full cooperation from the householder is essential since reliability of the data relies upon the entry. Therefore at times the data may not be authentic.

Weighing of Raw and Cooked Foods : This method is the most reliable one. But it is time-consuming and therefore it is not used for a large sample. As the investigator cannot leave the questionnaire different the householder she has to stay near and record the amount of various foods before and after cooking. The amount of food used can be recorded by weighing the leftovers after consumption. Often the housewife shows more raw foods to provide a wrong impression about their food intake. Therefore the investigator must have some practical knowledge about the quantities of cooked foods of diverse sources.

Procedure of Consumption: With a view to find out the nutritional adequacy of a family, a group adult consumption unit is calculated. In this method, the heterogenic groups of people i.e., school going, adult, geriatric, entire pregnant, a uniform calculation can be applied. This makes the total need calculation simple. For adult consumption unit a sedentary man is taken as one adult consumption unit (ACU) and the calorie needs of other age groups and special groups are expressed in terms of proportionate adult consumption units.

The adult consumption unit coefficient laid down by the Indian Council of Medical Research for the diverse groups and the kind of physical activity is mentioned below:

Adult male (Sedentary worker)		1.0
Adult male (Moderate worker)		1.2
Adult male (Heavy worker)		1.6
Adult female (Sedentary worker)		0.8
Adult female (Moderate worker)		0.9
Adult female (Heavy worker)		1.2
Adolescents	12 - 21 yrs	1.0
Children	9 - 12 yrs	0.8
Children	7 - 9 yrs	0.7
Children	5 - 7 yrs	0.6
Children	3 - 5 yrs	0.5
Children	1 - 3 yrs	0.4

Calculation of Nutritive Value of the Family Diet in Terms of Adult Consumption Unit: To find out the adult consumption unit of the family, the number of persons, their physical activities and their ages have to be recorded. Total consumption unit has to be calculated using the above table. For instance, for a family of 4, the adult consumption unit is provided below:

Member Age Occupation A.C.U.

Father	57 yrs	Retired Teacher	1.2
Mother	55 yrs	Housewife	0.9
Daughter.	23 yrs	Student	0.9
Son	17 yrs	Student	1.0
		Total	4.0

The adult consumption unit of the family is 4. The total calorie need of the family is calculated below.

The energy need for a sedentary = 2,800 kcals
man for 4 A.C.U. = 2,800 x 4 = 11, 200 kcals
Total cereals for the family per day = 460 gms x 4 = 1,040 gms

Actual Process

Food spoilage is brought about by the action of enzymes present in foods or because of the action of micro-organisms such as mould, yeast and bacteria or because of infestation with insects and worms. The environment unfavourable to the action of enzymes or to the growth of micro-organism is the main objective of food preservation. Natural and artificial methods are followed for this purpose. Preservatives are also added to foods to preserve them. diverse agents are used to bring either physical or chemical changes in food materials which are to be preserved.

As the principal spoilage agents are normally present in foods, destroying them or preventing their growth becomes the chief problem of food preservation. Any situation opposed to the growth of these organisms, whether by retarding their growth or by totally destroying them, aids in the preservation of food.

Methods commonly used to this end comprise general or cellar storage, refrigeration, canning, freezing, use of preservatives, drying and the exclusion of air. Although not a commonly used method at present, "cold sterilisation" or irradiation by beta and gamma rays to prolong the keeping quality of foods is being widely studied experimentally. The possibilities of this method of food preservation are great. When some of the major problems now limiting its use are solved, it is probable there will be radical changes in the present accepted methods of handling foods.

Refrigeration : Micro-organisms, although not easily destroyed by severe cold, are much less active at low temperature. Refrigeration is widely used both in homes and in commercial plants, as a means of maintaining the low temperature found satisfactory in the storage of perishable foods. Fresh milk, meat and similar foods are kept just above the freezing point.

Particular fruits and vegetables also keep better when cold, although as a rule they do not need low temperature for the limited time they are stored in the home as do the protein foods cited. Under-ripe fruits usually ripen rapidly at room temperatures. The keeping of cooked foods and leftovers is greatly facilitated by refrigeration, which is a more expensive but also a more efficient method of preservation than general storage.

Wrapping particular fruits and vegetables in paper or cellophane, or coating them with wax, improves the keeping quality. Pears and apples of high quality to be kept for winter use and oranges, lemons and grapefruit are often wrapped. The fruit or the paper in which it is wrapped may be treated to retard mould formation. Green peppers, tomatoes, oranges and other citrus fruits, cucumbers and cantaloupes are among the foods whose keeping quality is improved when they are coated with wax.

Heating: Food is generally preserved through the application of high temperatures. A temperature considerably above that of the body may result in outcome a pasteurised or a sterilised product. Foods are generally pasteurised by being held at a temperature of from 60° to 66° C, for thirty to forty minutes, during which time most of the organisms, although not all, are destroyed. Spores and some vegetative forms of bacteria remain, but as a rule those causing diseases are destroyed. This method is used for temporary preservation of milk, and for preservation of fruit juices and other fruit products of fragile flavour. Acid foods are rendered sterile by boiling for a comparatively short time.

Canning: If the effectiveness of pasteurisation and sterilisation is to extend over a period of time of practical significance, the material thus treated must be protected from fresh contamination by micro-organisms. The fact that in canning the sterilised or pasteurised food is afforded such. Protection accounts, at least in part, for the rapid growth in popularity of the process as a way of food preservation. The first foods preserved %9 canning were sealed in glass.

All prejudice which once existed against canned foods has been elimnated. Experiments have shown canned foods to be as wholesome as any form of preserved food. Generally the canned food is found to superior in food value to the freshly cooked. Canned foods are ready to use at a moment's notice. When food is to be kept over a long period of time, canned products are also most satisfactory and they do not need special temperatures or equipment for storage.

***Freezing* :** Freezing, like cold storage, does not destroy the micro-organisms enzymes present in the foods. It does render them more or less inactive however, so that frozen foods, when held at the proper temperature undergo change slowly. This applies to the nutritive value as well as to other chemical and physical characteristics.

Practically all general fruits are now preserved by freezing. However, the success with which the diverse fruits can be freezing varies. Most of the vegetables that are served cooked may be preserved by freezing. Scalding to destroy the enzymes is essential for successful preservation of almost all vegetables by this method. Hence vegetable to be used fresh and crisp in salads cannot be preserved by freezing. Fish, poultry and game, as well as beef, veal, pork and lamb to preserved by freezing. Frozen foods have little or no waste at the time using and little time is required for preparation. However, time must permit for the thawing of fruits to be served raw, and the cooking period of frozen meats and vegetables not already defrosted must be sufficient to allow for both thawing and cooking.

***Using Chemical Preservatives* :** Particular chemicals are helpful in preserving foods either by retarding preventing the growth of micro-organisms. These may be either added to the product or produced in it by fermentation. Sugar, a general he preservative, is often used in such quantities as to increase concentration of the food and make it an unfavourable medium for organisms. Salt and acid, in the form of vinegar or lemon juice, are substances added for their preservative action. Frequently, fermentation process is used to produce, inside the food itself, acetic lactic that exerts the same preservative action as is obtained by addition of vinegar. Salt is utilised as an aid in the preservation of vegetables by fermentation. It aids both in drawing out the juices are delaying the action of particular spoilage agents. Formerly, spices generally used in such quantities as to" have a definite preservative effect. Today the concentration in which they are used is usually less than employed earlier and it is believed that only a few of them are of signifcance as preservatives.

Particular preservatives such as borax, boric acid, sulfites, and formaldehyde, once sold as canning powders, are now considered injurious to human beings, and their use is usually prohibited.

Preserving and Pickling: Preserving and pickling are methods of preservation repeatedly employed. Long before canning was called a means of preserving foods, the utilisation of preservatives was practised. Salting of foods was an early method of preservation. Salt is a valuable preservative, both as an antiseptic and as an agent for removing water. Placing foods in brine of particular concentration promotes fermentation and the foods grow an agreeable flavour. Sugar in large amount is likewise a favourite preservative. Today, although practically every known variety of preserve and pickle is found in the market, the distinctive flavour of the homemade product is often preserved.

Drying: Removal of moisture is of benefit in preserving food. Although it is difficult to dry foods to the point of destroying micro-organisms, it is comparatively easy to dry foods so that no spoilage takes place. All organisms must have food on which to develop and grow and before the food present to them it must be in a solution which is rather dilute. These micro-organisms obtain food by osmotic action. Drying of foods to prevent spoilage does not essentially mean entire removal of the water, but it does mean concentration to such a point that the liquid is denser than the body fluid of the organisms. When the liquid outside the cell wall is more dense than that within, the liquid inside tends to be removed from the cell and the body processes are delayed or prevented. Hence, although the food is present it may not be present to the organism which might cause spoilage. Drying may retard enzyme action also, but in vegetables especially, the effect of enzymes is sufficient to cause comparatively rapid deterioration. Scalding to destroy the enzymes is highly advantageou in prolonging the storage life of dehydrated vegetables. Drying alters greatly the character of the food and needs some time for preparation both before and after the process.

Process of Priservation

The process of food preservation so far studies,such as application of heat,cold,removal of water and other methods utilised, have the common objective of decreasing the number of micro-organisms in food or holdingthem in check against further multiplication.In food preservation by fermentation,in contrast,multiplication of micro-organisms and their metabolic activities are encouraged.However,the organisms that are encouraged to grow and multiply are the select groups whose metabolic products assise food preservation.

Natural fermentations have play a important role in the preservation of foods from early times.Fruit juices exposed to air underwentnatural fermentation and received an alcoholic flavour.Milk on standing become slightly acidic.These changes were therefore used as a means of preservation of otherwise painful foods. At the same time the changes in the texure and taste of fermented foods were found desirable.These are the principal reasons for the continued use of fermentationin food processing and preservation.The list of foods produced by fermentation is highly long and comprises the following: cheese, curd, butter, all alcoholic beverages, pickles, sauerkraut, vinegar, bread, idli, soyasauce, coffee, tea and cocoa.

The term fermentation means the breakdown of carbohydrate material by micro-organisms(or enzymes) under anaerobic conditions. In common use, the term fermentation refers to both the anaerobic and aerobic breakdowns of carbohydrates and carbohydrate-like materials. Thus the conversation of lactose to lactic acid by streptococcus lactis which takes place under anaerobic situations and conversation of ethyl Alcohol to acetic acid by the bacteria acetobacter aceti under aerobic conditions are both reffered to as fermentation. The term is utilised in a still broader and less precise manner to mention fermented foods.

In the fermentation of foods, a complex mixture of carbohydrates, proteins, fats, etc., undergo modifications together with under the action of a number of micro-organisms and enzymes present. Thus, the carbohydrates and carbohydrate like materials undergo fermentation, proteinaceous material undergo proteolysis or "putrefactive" breakdown and lipids "lipolytic" breakdown. The nature and extent of these changes depend upon the food, sorts of micro-organisms present and conditions affecting their growth and metabolic pattern.

The preservative effect of fermentation is caused by the chemicals excreted by the micro-organism. The principal chemicals included are acids (particuarly lactic acid) and alcohol. These prohibit the growth or common pathogenic organisms in foods. The toxin producing clostridium botulinum, for instance, cannot grow at pH values below 4.6.

Benefits of fermentation: In addition to contributing to preservation, fermentation of foods has extra advantages. Fermentation produces flavour and textural changes and fermented foods are actually more nutricious than their unfermented counterparts. Microorganisms, in addition to breaking down complicated compounds, synthesize several vitamins and other growth factors. In fact, the industrial production of some vitamins is mainly by fermentation processes. The digestible carbohydrates and proteins of grains and seeds are enclosed in cellulosic and hemicellulosic structures which are indigestible materials. Fermentation, particularly by particular molds, breaks down the indigestible protective coating as they are rich in cellulose-splitting enzymes. The changed structure of the cells makes permeability of digestive juices easy. An extra advantage of such splitting of cellulose, hemicellulose and related materials is that such materials are changed into simpler sugars and sugar derivatives which increase the nutritional value of foods.

Microbial: Activities in Foods Microbial flora associated with foods have a broad complement of enzymes and, as already said, bring about fermentative, proteolytic and lipolytic changes among others. Proteolytic organisms break down proteins and other nitrogenous compounds resulting in putrid and rotten odour and flavor. Liplytic organisms hydrolyze lipids giving rise to free fatty acids resulting in rancidity. These changes are undesirable in foods. On the other hand, fermentative organisms changing carbohydrates and related compounds to alcohols, acids and carbon dioxide are not offensive to food tastes. When the concentration of their fermentative products increases, they control the growth of organisms of proteolytic and lipolytic types. However, fermentation of food is complex and only are sorts of changes is not possible. Thus, to receive a product of desired quality, controlled fermentation becomes esential.

Control of Fermentation in foods: The factors that influence the growth and metabolic activites of micro-organisms in food fermentation are the acidity, levels of alcohol, use of starters and levels of oxygen and salts.

Acid: Acids when present in food either as natural constituents, added or formed by fermentation, exert an prohibitory effect on harmful micro-organisms. However, the preservative effect of acid is lost if oxygen is available and surface molds develop which result in the fermentation of the acids. This happens during the ripening of cheddar cheese and constitutes a defect. Some yeasts tolerate moderately high acidity and produce ammonia from the breakdown of protein. This results in the neutralization of the acid present and permits the later growth of proteolytic and lipolytic bacteria as happens in the ripening of some sorts of cheese.

These sorts of changes also occur when raw milk is allowed to ferment. Raw milk contains many types of organisms. First, streptococcus lactis develops producing lactic acids and its growth is prohibited when the acid concentration reaches a particular level. Lactobacillus bacteria which are still more acid resistant then grow and produce more acid until the new level becomes prohibitory to their growth. This results in milk clotting and curd formation. In the high-acid environment, acid resistant yeasts and molds become active. The molds oxidize acid and the yeast produces alkaline end products. These result in a decrease of the acid level flavouring the development of proteolytic and lipolytic spoilage bacteria resulting in an off-odour and gassy condition in the curd.

Alcohol : Alcohol is also a preservative relying upon its concentration. Yeast converts sugar to alcohol but it cannot tolerate alcohol concentration of over 12 percent. When the alcohol content is below this level, as in beer, pasteurization of the food becomes essential to prevent spoilage.

Use of Starters: When a special type of micro-organisms is present in large numbers and is multiplying at a good rate, it keeps down the growth of other micro-organisms. This is the cause why in curd making or wine is added to the fresh milk or grape juice. However, in commercial food fermentation, pure cultures are utilised as starters to obtain products of required quality.

Temperature : Temperature has an effect on all microbial reactions and hence has to be controlled. In mixed fermentations, as in foods, control of temperature is all the more essential as different organisms have different temperature sensitivity. For instance, in the production of sauerkraut three major kind of organisms which grow well at different temperatures, are involved in fermentation. In sauerkraut fermentation, therefore, an beginning low temperature (about 21°C), which then is increased somewhat in the later stages of fermentation, is employed.

Oxygen: The level of oxygen conditions the extent of fermentation of some foods. Thus providing or removal of air or oxygen is used to encourage or prohibit a special micro-organism. Also, an organism may have diverse oxygen requirements for growth than it has for fermentation activity. For instance, bakers' yeast (S. cerevisiae) and wine yeast (S. ellipsoideus) both grow well and produce greater cell masses under aerobic conditions, but they ferment sugar more quickly under anaerobic condition. Thus, for production of bakers' yeast, the yeast is grown under the aerobic situation by bubbling air through the fermenting medium. In vinegar production there are two steps; the first step includes conversion of sugary substances to alcohol and the second the conversion of alcohol to acetic acid. Alcohol production is an anaerobic process and its conversion to acetic acid is an oxidative fermentation. Thus, in vinegar production, beginning fermentation should be anaerobic and this should be followed by generous supply of air.

Salt : Salt can also be used for controlling food fermentation. Lactic acid yielding micro-organisms utilised in the fermentation of pickles, sauerkraut, etc, are tolerant to salt solutions of the order of 10-18 percent. Many proteolytic and other spoilage organisms that can infest lactic bacteria are not tolerant to salt above 2.5 percent. Similarly, salt added to cheese curd controls proteolytic organisms during the long ripening periods.

Radiation : Preservation of food by ionizing radiations is a recently developed method but has not yet gained common acceptance. As is well known, the electromagnetic radiations suppress the growth of most micro-organisms. The feasibilities of employing nuclear radiation to sterilize food have been extensively studied since world war II. The harmful effects on the human body from radiations from nuclear explosions have given rise to suspicion in the minds of several people about the safety of the use of irradiated foods. Much work has been done on the safety and wholoesomeness of irradiated products but more careful investigations are needed before allowing the irradiated food to be used on a large scale.

Kinds of Ionizing Radiations: There are many types of radiant energy emitted from different sources which differ in wavelength, frequency, penetrating power and in the diverse effects they have upon biological systems. Irradiation of foods is carried out by exposing them to high-energy radiations which penetrate them and bring about changes within them. The principal radiations used in food irradiation are gamma rays (wavelengths<2A) emitted from excited radioactive elements ^{60}(CO) and accelerated negatively charged particles, electrons (b-particles), which are emitted from a hot cathode. Gamma rays are very resemble to x-rays, parring that they have much greater infiltrating power. An electron beam can be accelerated to very high speeds and increased energies so that it can penetrate foods, by passing it through electronic devices.

Ultraviolet radiations are also employed in preservation ; but they have a highly low degree of penetration and are employed to inactivate micro-organisms on the surface of food and to treat air, water and the surface of food processing equipment to reduce the number of micro-organisms.

Measurement of Radiation: Ionizing radiations are measured in terms of rads (and kilorad or megarad). A rad is

10 uj of energy absorbed by 1g of material. The radiation effects are concerned to dose by the relation:

$$n = n_0 \exp(-D/D_0)$$

where n = the number of live organisms following irradiation;

n_0 = initial number of organisms; D = does radiation received (rads),

and D_0 = a constant depending on sort of organism and environmental factors.

Different organisms are sensitive to irradiation to different extent as hinted below:

10^{3-4} - 10^7 rads – micro-organisms are killed

10^3 – 10^5 rads – insects are killed

10^3- 10^4 rads – sprouting of potatoes, onion, carrots, etc., are inhibited

10^2 – 10^3 rads – does is lethal for humans.

In the case of micro-organisms, the about sterilizing dose is: bacterial endospores, 3.0 x 10^6 rads; yeasts and fungi, -5.0 x 10^4 rads. There are, however, some micro-organisms with exceptional resistant up to 55 times greater dosage than gram-negative organisms like Escherichia coli.

Mode of Action of Radiations: When gamma rays and electron beam pass through foods, there are collisions between ionizing radiations and food particles at the atomic and molecular levels, Resulting in the production of ion pairs and free radicals. The reactions of these products among themselves and with other molecules result in physical and chemical phenomena inactivating micro-organisms in foods. Thus 'radiation of foods can be thought to be a means of achieving " cold sterilization" of food –a food is freed of micro-organisms with out the requirement for high temperatures.

Water is a unique ingredient in foods. Of all the reactions brought about by ionizing radiations,the one with water is the most important in yielding foods free of spoilage micro-organisms or path ogens or containing a greatlyh diminished content of spoilage organisms. Ionizing radiations split water to produce hydrated electrons (e'_{aq})' very reactive hydrogen (H) and hydroxyl (OH)radicals,exiced water (H2O) and ionized water molecules (h2O).Interactions between these products,or these products and other components of food comprising oxygen result in the formation of highly reactive species like hydrogen peroxide.

Hydrogen peroxide is a strong oxiditing agent and a toxicant to micro-organisms.

Uses of Radiations Ionizing radiations may be utilised for sterilization of foods in hermetically sealed packs, reduction in the size of the spoilage flora on painful foods, elimination of pathogens in foods, control of infestation in stored cereals, prohibition of sprouting of potatoes, onions, etc. and retardation of the development of peckled mushrooms . The destruction of the growing point causing tissue darkening of the slproouting areas of vegetables such as potatoes by radiations, eliminates the risk of sprouting during storage.

Irradiation does not very much affect the nutritional properties of foods. The destruction of diverse amount of nutrients are of the same degree as in heat processing. Ionizing radiations can also hydrolyze and modify proteins , starch and cellulose. Thus, irradiation can help to improve nutritive value of particular plant materials. Important levels of toxic carcinogenic substances are not produced in foods irradiated with sufficient dose of radiations. At proper dose of radiations, sterilized or pasteurized foods, are safe from microbial stand point.

Irradiation of foods can also result in particular undesirable effects. Ionizing radiation, in excessive doses, can alter the structure of organic and biochemical compounds in food resulting in food damage. Degradation of carbohydrates may result in loss of texture and colour. Protein degradation results in unde rsirable lchanges in colour and odour and brings about liquefaction.Egg-white proteins, on irradiation, become thin and watery. Irradiation of fats results in the development of off-flavour and off-odour and the loss of the common antioxidants. Foods irradiated with large doses of radiation might themselves become radioactive and prove harmful when concurred. In the irradiation of foods for preservation, the control of the radiation process is very essential. The radiation dose must be carefully controlled to destroy micro-organisms and inactivate many food enzymes. It should be adequale to destroy the pathogenic and spoilage causing orgainms .Spart from the intensity of radiation areto be controlled. He longer the food is in th radiation path, the more it will bsorb, adiation energy must be provided in such a manner that it reaches every particles of food to to make sure substantial killing of all micro-organisms.

As already said, the use of ionizing radiatons for food preservation has beeb followed very slowly. This is largely because of the unacceptable flavour of some foods irradiated to sterilization; stage. Also the fear that radioactivity might be induced in a food has come in the way of its extended use. However, recent reappraisal has resulted in the recommendation that irradiation dose up to 1M rad is not hazardous.

The disinfection of grains, the preservation, the prevention of sprouting of potatoes and other vegetable are of great economic value in less advanced countries lacking refrigeration facilities, losses from insects and spoilage of food account for more than half the total food production. In such cases radiation of foods will be of great importance. This requires radiation facilities at a cost that third world countries can afford.

□□□